Inclusion Metallurgy

Inclusion Metallurgy

Editors

Yanling Zhang
Guoguang Cheng
Zhonghua Zhan

MDPI • Basel • Beijing • Wuhan • Barcelona • Belgrade • Manchester • Tokyo • Cluj • Tianjin

Editors

Yanling Zhang
State Key Laboratory of Advanced Metallurgy, University of Science and Technology Beijing, Beijing, China

Guoguang Cheng
State Key Laboratory of Advanced Metallurgy, University of Science and Technology Beijing, Beijing, China

Zhonghua Zhan
State Key Laboratory of Advanced Metallurgy, University of Science and Technology Beijing, Beijing, China

Editorial Office

MDPI
St. Alban-Anlage 66
4052 Basel, Switzerland

This is a reprint of articles from the Special Issue published online in the open access journal *Metals* (ISSN 2075-4701) (available at: https://www.mdpi.com/journal/metals/special_issues/Inclusion_Metal).

For citation purposes, cite each article independently as indicated on the article page online and as indicated below:

LastName, A.A.; LastName, B.B.; LastName, C.C. Article Title. *Journal Name* **Year**, *Volume Number*, Page Range.

ISBN 978-3-0365-8446-1 (Hbk)
ISBN 978-3-0365-8447-8 (PDF)

Contents

metals MDPI

Editorial

Inclusion Metallurgy

Yanling Zhang *, Guoguang Cheng and Zhonghua Zhan

State Key Laboratory of Advanced Metallurgy, University of Science and Technology Beijing, Beijing 100083, China
* Correspondence: ustbzly1108@163.com; Tel.: +86-010-8237-5191

1. Introduction and Scope

Non-metallic inclusions have a great influence on the cleanliness and mechanical properties of steel. By controlling the size and composition of inclusions, the excellent properties of "clean steel" can be maintained. At the same time, in terms of our understanding of inclusions' behavior using thermodynamics principles, the design and control of the composition, shape, size, and distribution of non-metallic inclusions in different steels can significantly enhance their properties.

The primary focus of this Special Issue is on recent advancements in inclusion engineering which have the aim of controlling steel cleanliness and microstructure through modeling and experimental work. Research into the particularly interesting theme of the formation mechanism and evolution control methods of inclusions in the smelting process in laboratories and steel plants was welcomed. The study of the agglomeration and floatation of inclusions and the kinetics of slag adsorption in the process of refining and solidification were also potential themes.

2. Contributions

In this Special Issue, 10 high-quality papers covering a wide range of inclusion metallurgy research, including rare earth treatment, electroslag remelting, the formation and evaluation of inclusions during steel smelting, and the effect of inclusions on mechanical properties and solidification structures have been published.

Five papers focused on the effect of rare earth on steel, which clarified the effect of sulfur on inclusions and the mechanical properties of Ce-Mg-treated, resulfurized SCr420H steel [1]; the evolution of inclusions in austenitic heat-resistant steel with different levels of Ce content during protective argon gas atmosphere electroslag remelting (ESR) [2]; the effect of Ce on the morphology of manganese sulfide [3]; the effects of rare earth La–Ce alloying treatment on the characteristics of steel [4]; and the effect of rare earth Ce content on the morphology, composition, type, and size distribution of inclusions in W350 non-oriented silicon steel [5].

Four papers focused on the formation and evolution of inclusions during steel smelting, which covered the formation and removal mechanisms of inclusions in 42CrMo4 steel during the steelmaking process [6]; the evolution of inclusions and the control strategies used to improve the cleanliness of molten steel in Ti-bearing IF steel [7]; the source and formation of magnetic particle inspection defects identified on the near-surface of the Cr5 back-up roll-forged ingot [8]; and the origin, evolution, and formation mechanism of MgO-based inclusions in Si-Mn-killed steel [9].

One paper focused on the solidification behavior and structure of 2311 die steel with a cross-section dimension of 415 × 2270 mm at different casting speeds, specific water flow, and superheat [10].

3. Conclusions and Outlook

Topics such as the effect of rare earth on steel, the formation and evolution of inclusions during steel smelting, and solidification behavior and structure are covered in this Special

Citation: Zhang, Y.; Cheng, G.; Zhan, Z. Inclusion Metallurgy. *Metals* **2023**, *12*, 827. https://doi.org/10.3390/met13050827

Received: 28 March 2023
Accepted: 31 March 2023
Published: 23 April 2023

Issue, which presents the latest developments in inclusion metallurgy research and their applications. As Guest Editors of this Special Issue, we hope that the reported studies will be useful to researchers in advancing their respective research areas.

Funding: This work was supported by the National Natural Science Foundation of China, grant numbers U196021 and 51874034, the National Natural Science Foundation of China N. 52074075 and the National Key Research and Development Program of China N. 2016YFB0300105, the National Natural Science Foundation of China (Grant Nos. 52074027 and 51874026), the National Natural Science Foundation of China, grant number 51874028, the National Natural Science Foundation of China (Grant No. 51774031), the National Natural Science Foundation of China, grant number 51774030, the National Natural Science Foundation of China (Nos.52274312 and 51804003), the Fund of Education Department of Anhui Province (Nos. 2022AH050291, 2022AH050293), the Open Project Program of Anhui Province Key Laboratory of Metallurgical Engineering & Resources Recycling (Anhui University of Technology) (Nos. SKF21-04 and SKF21-05), the University Natural Science Research Project of Education Department of Anhui Province (No. KJ2021A0396), and the Jiangxi Province Major Scientific and Technological Research and Development Special Funding Project (20213AAE01009).

Acknowledgments: As Guest Editors, we highly appreciate the valuable research from the contributing authors, the professionalism of the reviewers and editors, and the efforts of the staff working on this Special Issue. Special gratitude goes to the *Metals* Editorial Office for its great support.

Conflicts of Interest: The authors declare no conflict of interest.

References

1. Sun, M.; Jiang, Z.; Li, Y.; Chen, C.; Ma, S.; Ji, Y.; Wang, J.; Liu, H. Effect of Sulfur Content on the Inclusion and Mechanical Properties in Ce-Mg Treated Resulfurized SCr420H Steel. *Metals* **2022**, *12*, 136. [CrossRef]
2. Wang, Z.; Shi, C.; Wang, S.; Li, J.; Zhu, X. Evolution and Formation of Non-Metallic Inclusions during Electroslag Remelting of Ce-Bearing 15Cr-22Ni-1Nb Austenitic Heat-Resistant Steel. *Metals* **2022**, *12*, 2094. [CrossRef]
3. Zhuo, C.; Liu, R.; Zhao, Z.; Zhang, Y.; Hao, X.; Wu, H.; Sun, Y. Effect of Rare Earth Cerium Content on Manganese Sulfide in U75V Heavy Rail Steel. *Metals* **2022**, *12*, 1012. [CrossRef]
4. Wang, H.; Niu, Y.; Ling, H.; Qiao, J.; Zhang, Y.; Zhong, W.; Qiu, S. Effects of Rare Earth La–Ce Alloying Treatment on Modification of Inclusions and Magnetic Properties of W350 Non-Oriented Silicon Steel. *Metals* **2023**, *13*, 626. [CrossRef]
5. Wang, H.; Niu, Y.; Ling, H.; Qiao, J.; Zhang, Y.; Zhong, W.; Qiu, S. Modification of Rare Earth Ce on Inclusions in W350 Non-Oriented Silicon Steel. *Metals* **2023**, *13*, 453. [CrossRef]
6. Qiao, T.; Cheng, G.; Huang, Y.; Li, Y.; Zhang, Y.; Li, Z. Formation and Removal Mechanism of Nonmetallic Inclusions in 42CrMo4 Steel during the Steelmaking Process. *Metals* **2022**, *12*, 1505. [CrossRef]
7. Yuan, B.; Liu, J.; Zeng, J.; Zhang, M.; Huang, J.; Yang, X. Evolution of Inclusions and Cleanliness in Ti-Bearing IF Steel Produced via the BOF–LF–RH–CC Process. *Metals* **2022**, *12*, 434. [CrossRef]
8. Zhang, W.; Wang, G.; Zhang, Y.; Cheng, G.; Zhan, Z. Formation Mechanism and Improvement of Magnetic Particle Inspection Defects in Cr5 Backup Roller Forged Ingot. *Metals* **2022**, *12*, 295. [CrossRef]
9. Zhao, J.; Chu, J.; Liu, X.; Wang, M.; Cai, X.; Ma, H.; Bao, Y. Source and Transformation of MgO-Based Inclusions in Si-Mn-Killed Steel with Lime-Silicate Slag. *Metals* **2022**, *12*, 1323. [CrossRef]
10. Xu, L.; Zhang, P.; Shuai, Y.; Shi, P.; Zhan, Z.; Wang, M. Study of Process Parameters on Solidification Structure and Centre Grain Size of 2311 in 420 Mm Extra-Thick Continuously Cast Slabs. *Metals* **2023**, *13*, 47. [CrossRef]

metals MDPI

Article

Evolution and Formation of Non-Metallic Inclusions during Electroslag Remelting of Ce-Bearing 15Cr-22Ni-1Nb Austenitic Heat-Resistant Steel

Zhongwei Wang, Chengbin Shi *, Shijun Wang, Jing Li and Xin Zhu

State Key Laboratory of Advanced Metallurgy, University of Science and Technology Beijing (USTB), Beijing 100083, China
* Correspondence: chengbin.shi@ustb.edu.cn

Abstract: The evolution of inclusions in austenitic heat-resistant steel with different Ce content during protective argon gas atmosphere electroslag remelting (ESR) was studied. All oxide inclusions in the Ce-free consumable electrode are $MgO{\cdot}Al_2O_3$. A part of these $MgO{\cdot}Al_2O_3$ inclusions was removed before metal droplets entered the liquid metal pool during the ESR. The soluble oxygen (arising from the reoxidation) reacted with soluble aluminum, calcium, and magnesium in liquid steel to form $MgO{\cdot}Al_2O_3$ and $CaO–Al_2O_3$ inclusions in liquid steel. All oxide inclusions in the electrode with 0.016 mass% Ce are Ce_2O_2S. A portion of these Ce_2O_2S inclusions was dissociated into soluble oxygen, cerium, and sulfur in liquid steel during the ESR process, whereas the others were removed by absorbing them into molten slag. The oxide inclusions in the liquid metal pool and remelted ingot were Ce_2O_3, $CeAlO_3$, and Ce_2O_2S. The $CeAlO_3$ and Ce_2O_3 inclusions were reoxidation products formed by the chemical reaction between the soluble oxygen, soluble aluminum, and cerium. The oxide inclusions in the electrode with 0.300 mass% Ce are CeS and Ce_2O_2S. These CeS inclusions were removed by molten slag adsorption during the ESR. A part of these Ce_2O_2S inclusions was removed by slag adsorption, and the remaining entered into the liquid metal pool. The oxide inclusions in the liquid metal pool and the ingot were Ce_2O_3 and Ce_2O_2S. The Ce_2O_3 inclusions were formed through the chemical reaction between the soluble oxygen and cerium in the liquid metal pool. The Ce_2O_2S inclusions in the liquid pool originate from reoxidation products during the ESR process and the relics from the electrode.

Keywords: inclusion; cerium; electroslag remelting; reoxidation; evolution

Citation: Wang, Z.; Shi, C.; Wang, S.; Li, J.; Zhu, X. Evolution and Formation of Non-Metallic Inclusions during Electroslag Remelting of Ce-Bearing 15Cr-22Ni-1Nb Austenitic Heat-Resistant Steel. *Metals* **2022**, *12*, 2094. https://doi.org/10.3390/met12122094

Academic Editors: Mark E. Schlesinger and Antoni Roca

Received: 28 October 2022
Accepted: 5 December 2022
Published: 6 December 2022

1. Introduction

Austenitic heat-resistant steel of Ni-Cr system is extensively utilized in automotive engines and fossil fuel power generation plants applications at moderately elevated temperatures, such as fasteners and bolts, and heat exchangers, because of their superior strength, toughness, oxidation resistance, and organizational stability [1–3]. Except the advantages of some types of fine non-metallic inclusions through acting as the nucleation sites to generate intragranular acicular ferrites and the pinning particles to inhibit the growth of austenite grains [4], inclusions are generally detrimental to mechanical properties of steel, such as creep strength, fatigue strength, corrosion resistance, tensile properties, toughness, and ductility [5–10]. The local stress concentration induced by large inclusions often causes cracking at inclusion-steel matrix interface during working of steel [11]. The detriment degree of inclusions is strongly dependent on their number density, size, and chemical compositions.

Rare earth Ce can improve the corrosion resistance [12,13], oxidation resistance [14,15], grain refinement [16,17] and reduction of dendritic spacing [18,19] of heat-resistant steels and improve the mechanical properties of austenitic heat-resistant steels [20,21]. The generation of Ce-bearing non-metallic inclusions in Ce-containing steel was unavoidable. Ce-containing inclusions were detrimental to the fatigue strength, toughness and ductility, impact strength, tensile properties, corrosion resistance, initiation of fatigue cracking, and stress concentration cracking of steel [22–24]. Many studies on the evolution of Ce-bearing inclusions have been conducted in industrial-scale secondary refining (LF and RH refining) or laboratory-scale crucible experiments of slag-steel reactions [25,26]. However, no studies have been reported on the evolution of Ce-bearing inclusions in the ESR process.

In the current study, the evolution of inclusions in the steel with different Ce contents during ESR was investigated. The effect of reoxidation during the ESR on inclusion evolution was ascertained. The evolution mechanism of Ce-bearing inclusions during the ESR was elucidated.

2. Experimental

2.1. ESR Experimental Procedure

The chemical compositions of the consumable electrodes are shown in Table 1. The oxide scale on the electrode steel surface was thoroughly removed before ESR experiments. The pre-melted slag (40.92 mass% CaF_2, 27.09 mass% CaO, 0.52 mass% MgO, 29.02 mass% Al_2O_3, 0.92 mass% SiO_2, 0.21 mass% FeO) was roasted at 700 °C (973 K) for 8 h to remove moisture before the ESR trials.

Table 1. Chemical compositions of the consumable electrodes (mass%).

Trials No.	C	Si	Mn	Mo	Ti	Cr	Ni	Al	Nb	S	Ca	Ce	Mg	N	O
T1	0.02	0.5	0.83	2.49	0.18	15.1	22	0.15	0.98	0.0032	0.0002	0	0.0004	0.0071	0.0010
T2	0.02	0.5	0.89	2.53	0.14	14.9	22	0.10	0.99	0.0009	0.0002	0.016	0.0003	0.0120	0.0006
T3	0.02	0.5	0.78	2.50	0.21	14.6	22	0.15	1.03	0.0025	0.0002	0.300	0.0003	0.0031	0.0007

Three ESR trials were conducted in protective argon gas atmosphere. The operating current, voltage, and outlet temperature of the mold cooling water were maintained at about 2500 A, 32 V, and 313 k. During electroslag remelting of the steel electrode with 0.016 mass% Ce, Al shots were added for deoxidation (addition amount: 1 kg/ton of steel). For deoxidation in ESR trial T3, Al shots were added (2 kg/ton of steel) during the ESR of the electrode with 0.300 mass% Ce. A steel sample was taken from the liquid metal pool during the ESR refining using a vacuum sampling tube made of quartz, followed by quenching in water. The remelted ingots corresponding to the ESR trials T1, T2, and T3 were designated as ingots C1, C2, and C3, respectively.

2.2. Compositional Analysis and Inclusions Characterization

The steel samples were taken from the consumable electrodes and the remelted ingots of the ESR for chemical composition analysis. The total oxygen (composed of both the soluble oxygen and the oxygen bonded as oxide inclusions) content of the steel was measured by the inert gas fusion-infrared absorptiometry. The sulfur and carbon contents of the steel were measured by the combustion-infrared absorption technique. The nitrogen content was determined by the inert gas fusion-thermal conductivity method. The contents of soluble aluminum, total calcium, magnesium, silicon, cerium, and manganese in the electrodes and ingots were measured by inductively coupled plasma atomic emission spectroscopy (ICP-AES).

Steel samples for metallographic observation were taken the consumable electrodes and the mid-height of each remelted ingot at the mid-radius position. To reveal the evolution of inclusions in the ESR process, steel samples collected from the liquid metal pool with vacuum sampling tubes were polished after mechanical grinding. Inclusions exposed on the cross-section of the polished sample were analyzed in terms of their chemistry, size, and morphology by scanning electron microscope (SEM; FEI Quanta-250, FEI Corp, Hillsboro, OR, USA) equipped with an energy-dispersive X-Ray spectrometer (EDS, XFlash 5030; Bruker, Karlsruhe, Germany).

Furthermore, the size of inclusions on the polished sample cross-section was calculated using the equivalent circle diameter (ECD). The equivalent circle diameter of the inclusions smaller than 1 μm is not included in analysis.

3. Results and Discussion

3.1. Steel Composition

The chemical compositions of remelted ingots are shown in Table 2. The Ce contents of the remelted ingots C2 and C3 are 0.0055 mass% and 0.063 mass%, respectively. The yields of Ce during the ESR are 34.4% and 21.0%, respectively.

Table 2. Chemical Compositions of Remelted Ingots (mass%).

Ingot No.	C	Si	Mn	Mo	Ti	Cr	Ni	Al	Nb	S	Ca	Ce	Mg	N	O
C1	0.02	0.5	0.82	2.60	0.13	15.5	22	0.19	1.00	0.0007	0.0002	0	0.0004	0.0076	0.0016
C2	0.02	0.5	0.89	2.60	0.12	15.4	22	0.23	1.00	0.0007	0.0002	0.0055	0.0003	0.0140	0.0013
C3	0.02	0.5	0.83	2.50	0.15	14.6	22	0.15	1.02	0.0016	0.0002	0.0630	0.0003	0.0054	0.0018

The aluminum content of the steel increases from 0.15 mass% in the Ce-free electrode to 0.19 mass% in the remelted ingot C1. The aluminum content of steel in the ESR trial T2 increases from 0.10 mass% in the consumable electrode to 0.23 mass% in the ingot C2. The aluminum content of the steel in the ESR trial T3.

The slag-steel reactions during the ESR are a potential source of soluble aluminum pickup in liquid steel. The chemical reaction between Ce in liquid steel and Al_2O_3 in the slag can be expressed as follows [27]:

$$2[Ce] + (Al_2O_3)_{slag} = (Ce_2O_3)_{slag} + 2[Al] \tag{1}$$

$$\Delta G^{o} = 205703 + 34T \ (J/mol) \tag{2}$$

$$K = \frac{a_{Al}^2 \cdot a_{Ce_2O_3}}{a_{Ce}^2 \cdot a_{Al_2O_3}} = \frac{(f_{Al}[\%Al])^2 \cdot a_{Ce_2O_3}}{(f_{Ce}[\%Ce])^2 \cdot a_{Al_2O_3}} \tag{3}$$

where $a_{Ce_2O_3}$ and $a_{Al_2O_3}$ are the activities of Ce_2O_3 and Al_2O_3 in the slag, respectively. f_{Al} and f_{Ce} are the activity coefficients of soluble aluminum and Ce in the liquid alloy, respectively, which can be calculated by the following equation [28].

$$\lg f_i = \sum e_i^j [\%j] \tag{4}$$

where e_i^j is the first-order interaction parameter.

The activities of the Al_2O_3 and Ce_2O_3 relative to pure solid standard states in the slag melts at 1873 K (1600 °C) are estimated with FactSage 8.0 (FToxid database; ThermFact/CRCT, Montreal, Canada). According to the chemical composition of the electrode with 0.300 mass% Ce combined with the first-order interaction parameters (as shown in Table 3), the Gibbs free energy change for the chemical reaction (1) was calculated to be 312.89 kJ/mol. It suggests that Ce in the electrode with 0.300 mass% Ce cannot reduce the Al_2O_3 in the slag during the ESR process, leading to the Al pickup in the steel. The Al pickup in liquid steel is caused by the Al addition in the ESR process.

Table 3. First-order interaction parameters used in the current study [29–32].

	C	Si	Mn	Mo	Ti	Cr	Ni	Al	Mg	Ca	S	Ce	N	O
O	−0.421	−0.066	−0.021	0.005	−1.8	−0.033	0.006	−1.17	−300	-	−0.133	−0.57	0.057	−0.17
S	0.111	0.075	−0.026	-	−0.27	−0.0105	-	0.041	−1.82	−110	−0.046	−1.91	0.01	−0.27
Al	0.091	0.056	0.035	-	0.004	-	−0.029	0.045	−0.3	−0.047	0.035	−0.52	−0.057	−1.98
Ce	−0.077	-	-	-	−3.62	-	-	−2.25	-	-	−8.36	−0.003	−6.612	−5.03

In the present study, the oxygen contents of the electrodes with different Ce contents apparently increase during the ESR to 0.0016 mass%, 0.0013 mass%, and 0.0018 mass% in ingots C1, C2, and C3, respectively. It indicates that the reoxidation of the liquid steel occurred in these three ESR trials. The sulfur contents of steel decrease during the ESR to 0.0007 mass%, 0.0007 mass%, and 0.0016 mass% in ingots C1, C2, and C3, respectively.

The difference in the oxygen potential between liquid steel with low oxygen content and the molten slag contributes to soluble oxygen pickup in liquid steel during the protective argon gas atmosphere ESR [33]. The introduction of FeO (newly formed FeO during ESR and the unremoved oxide layer on the surface of the electrode steel) to the molten slag pool during ESR is difficult to be prevented, and it is the source of the increased oxygen content of liquid steel according to the chemical reaction $(FeO)_{slag}$ = [Fe] + [O] ($\Delta G^o = -139{,}000 + 57.1T$ (J/mol)) [34].

The interaction coefficients used in the present calculation are shown in Table 3. The Gibbs free energy change for the chemical reaction $(FeO)_{slag}$ = [Fe] + [O] for the ESR trials T1, T2, and T3 was calculated to be −102.75 kJ/mL, −105.99 kJ/mL, and −100.92 kJ/mL, respectively. It indicates that FeO in molten slag transfers oxygen to liquid steel, resulting in the reoxidation of liquid steel during ESR. Meanwhile, the desulfurization reaction (CaO) + [S] = $(CaS)_{slag}$ + [O] is also an important source of soluble oxygen pickup in the liquid steel during the ESR [35].

3.2. Inclusions in Consumable Steel Electrode

The SEM images and EDS spectra of typical inclusions in the consumable electrodes are shown in Figures 1–5. As shown in Figure 1, the typical inclusions in the Ce-free electrode are (Ti,Nb)N, $MgO{\cdot}Al_2O_3$ inclusions surrounded by (Ti,Nb)N and $MgO{\cdot}Al_2O_3$ inclusions, respectively. Figure 1a shows typical (Ti,Nb)N inclusions in the electrode with near-square shape. Figure 1c,d show typical $MgO{\cdot}Al_2O_3$ inclusions in Ce-free electrode, and the morphology of these spinel inclusions is nearly spherical. The $MgO{\cdot}Al_2O_3$ spinel inclusion plays a nucleation role for (Ti,Nb)N.

Figures 2 and 3 show the typical inclusions in the electrode with 0.016 mass% Ce. The observed inclusions are (Ti,Nb)N, and Ce_2O_2S, respectively. Figure 2a shows typical inclusions (Ti,Nb)N in the consumable electrode with a clear angular shape. Figure 2b,c show the typical inclusions of Ce_2O_2S in the consumable electrode, showing both near spherical or irregular morphologies.

The typical inclusions in the electrode with 0.300 mass% Ce are shown in Figures 4 and 5. The rare-earth inclusions in the electrode are identified as CeS and Ce_2O_2S, respectively. It can be seen from SEM photographs that these inclusions are near-spherical. It is found from ESD element mappings of the Ce_2O_2S inclusion that the elements are uniformly distributed on cross-section of the inclusions.

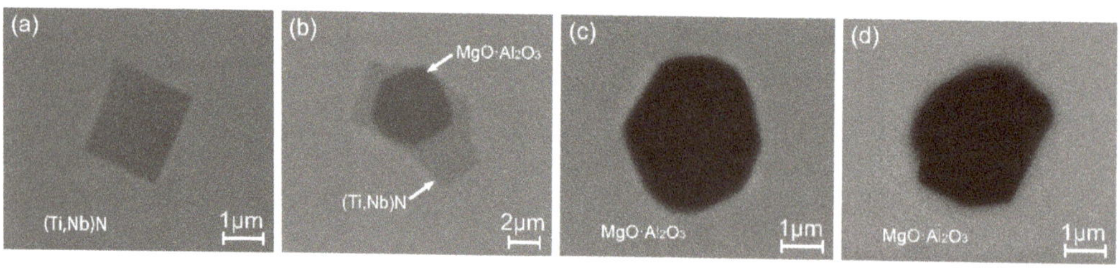

Figure 1. Typical inclusions in the Ce-free electrode. (**a**) (Ti,Nb)N, (**b**) $MgO{\cdot}Al_2O_3$+ (Ti,Nb)N, (**c**) and (**d**) $MgO{\cdot}Al_2O_3$.

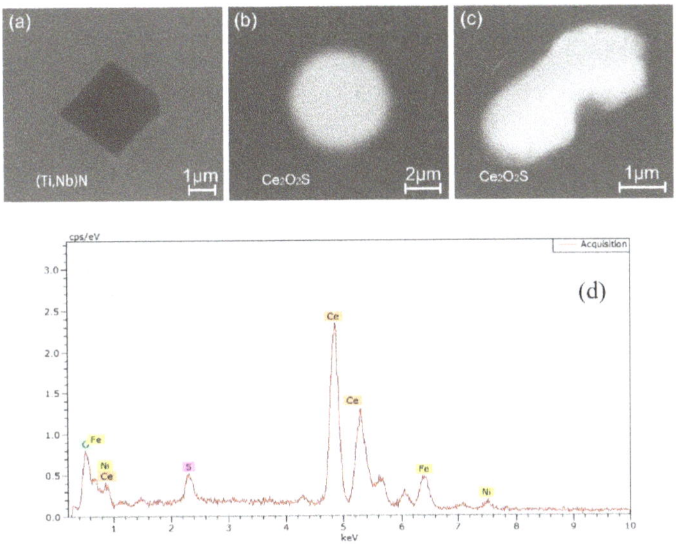

Figure 2. SEM images and EDS spectra of typical inclusions in the electrode with 0.016 mass% Ce. (**a**) (Ti,Nb)N, (**b**,**c**) Ce_2O_2S. (EDS spectrum shown in (**d**) corresponds to the inclusion shown in (**c**)).

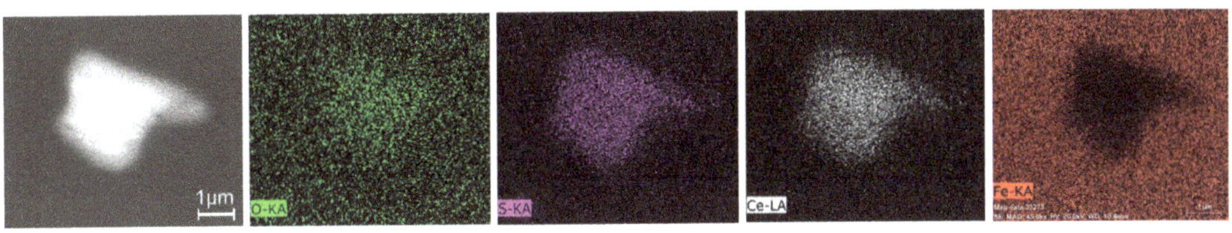

Figure 3. EDS element mappings of Ce_2O_2S inclusion in the electrode with 0.016 mass% Ce.

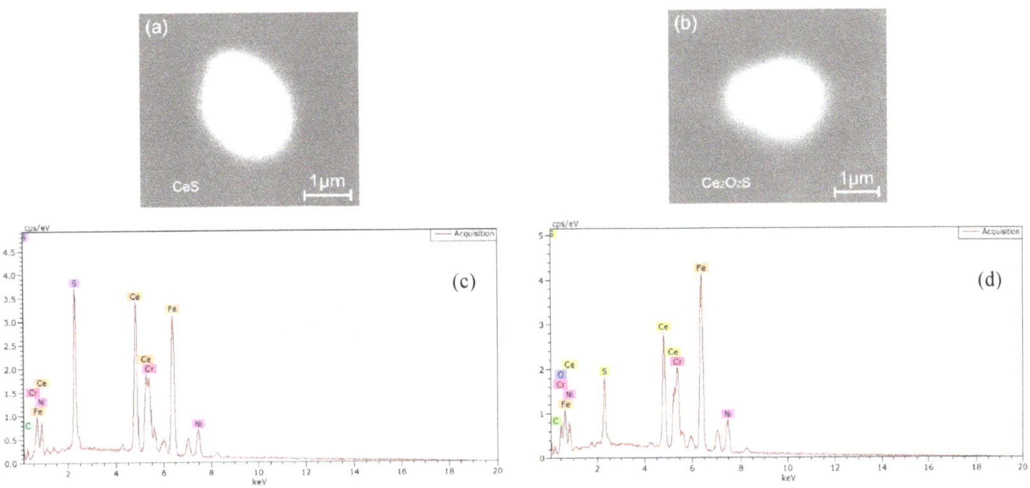

Figure 4. SEM images and EDS spectra of typical inclusions in the electrode with 0.300 mass% Ce. (**a**) CeS, (**b**) Ce_2O_2S. (EDS spectra shown in (**c**) and (**d**) correspond to the inclusions shown in (**a**) and (**b**), respectively).

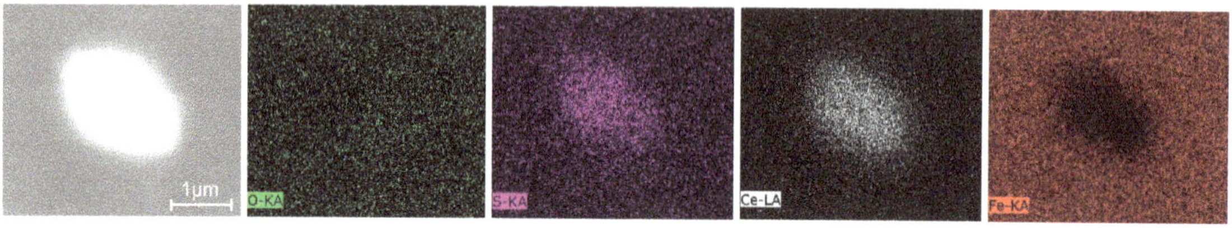

Figure 5. EDS element mappings of Ce_2O_2S inclusion in the electrode with 0.300 mass% Ce.

3.3. Inclusions in the Liquid Metal Pool and Remelted Ingots

The SEM micrographs, EDS spectra, and elemental mappings of typical inclusions observed in the liquid metal pool during the ESR trial T1 and remelted ingot C1 are shown in Figures 6–9. The inclusions in the liquid metal pool of the ESR trial T1 are (Ti,Nb)N, $MgO·Al_2O_3$ inclusions wrapped by (Ti,Nb)N, $CaO–Al_2O_3$, and $MgO·Al_2O_3$.

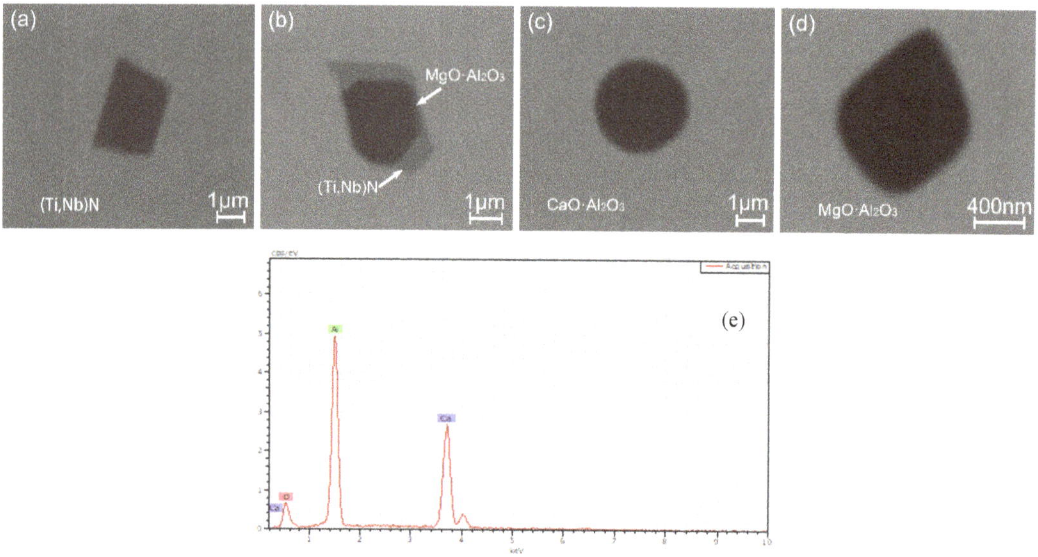

Figure 6. SEM images and EDS spectra of inclusions in the liquid metal pool during ESR trial T1. (**a**) (Ti,Nb)N, (**b**) $MgO·Al_2O_3$ + (Ti,Nb)N, (**c**) $CaO–Al_2O_3$, (**d**) $MgO·Al_2O_3$. (EDS spectrum shown in (**e**) corresponds to the inclusion shown in (**c**)).

Figure 7. EDS element mappings of $CaO–Al_2O_3$ inclusion in the liquid metal pool during the ESR trial T1.

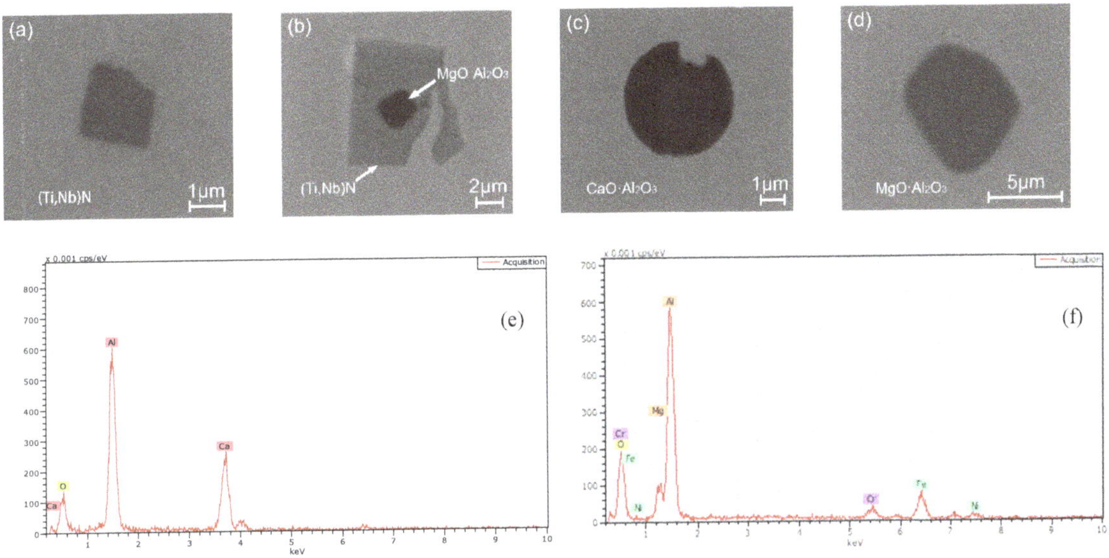

Figure 8. SEM images and EDS spectra of typical inclusions observed in ingot C1. (**a**) (Ti,Nb)N, (**b**) $MgO{\cdot}Al_2O_3$ + (Ti,Nb)N, (**c**) $CaO–Al_2O_3$, (**d**) $MgO{\cdot}Al_2O_3$. (EDS spectra shown in (**e**) and (**f**) correspond to the inclusions shown in (**c**) and (**d**), respectively).

Figure 9. EDS element mappings of $CaO–Al_2O_3$ inclusion in the ingot C1.

Four types of inclusions are found in ingot C1, namely (Ti,Nb)N inclusions, $MgO{\cdot}Al_2O_3$ inclusions wrapped by (Ti,Nb)N, $CaO–Al_2O_3$, and $MgO{\cdot}Al_2O_3$. Figures 7 and 9 show the EDS element mappings of $CaO–Al_2O_3$ inclusions in the liquid metal pool of the ESR trial T1 and remelted ingot C1, respectively. CaO and Al_2O_3 are uniformly distributed over the entire cross-section of the inclusion.

SEM micrographs, EDS spectra, and elemental mappings of typical inclusions in the liquid metal pool of ESR trial T2 and remelted ingot are shown in Figures 10–13. The types of inclusions in the liquid metal pool of the ESR trial T2 are (Ti,Nb)N, Ce_2O_2S, and $CeAlO_3$ encapsulated by (Ti,Nb)N, $CeAlO_3$, and Ce_2O_3. The types of inclusions are (Ti,Nb)N, Ce_2O_2S, $CeAlO_3$ surrounded by (Ti,Nb)N, and Ce_2O_3. It is observed that the types of inclusions in the remelted ingot did not change compared to the types of inclusions in the liquid metal pool.

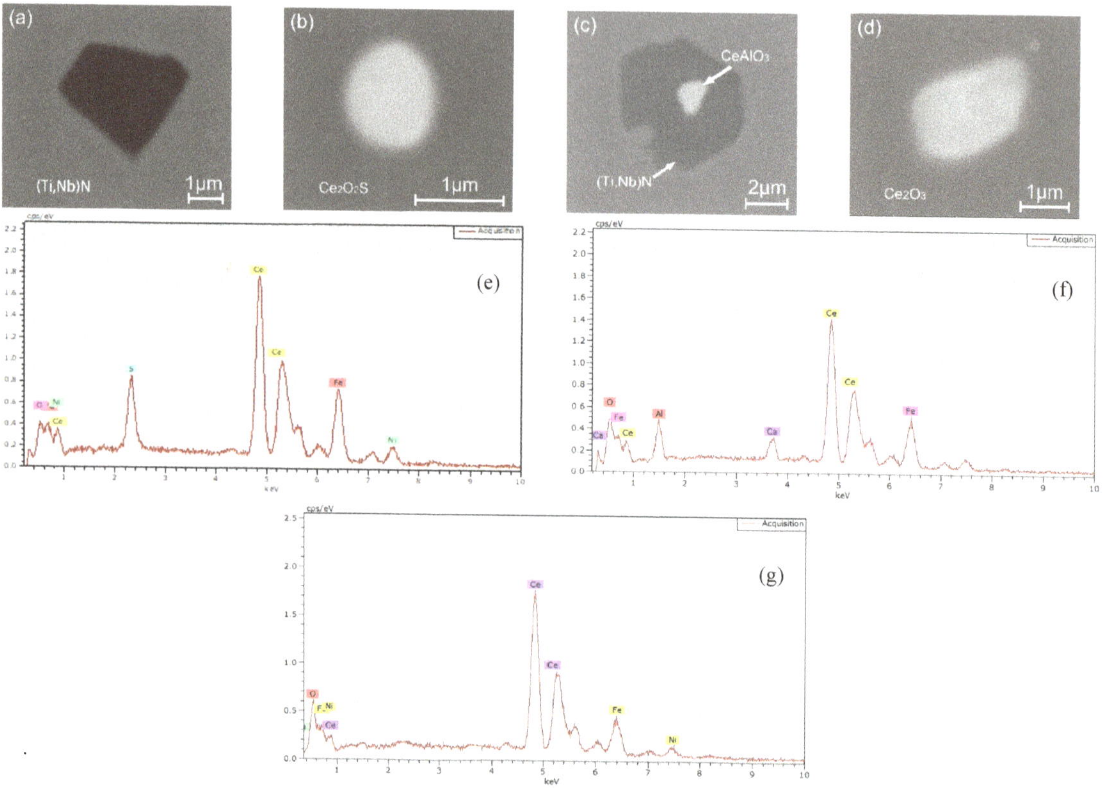

Figure 10. SEM images and EDS spectra of typical inclusions in the liquid metal pool during the ESR trial T2. (**a**) (Ti,Nb)N, (**b**) Ce_2O_2S, (**c**) $CeAlO_3$ + (Ti,Nb)N, (**d**) Ce_2O_3. (EDS spectra shown in (**e**), (**f**) and (**g**) correspond to the inclusions shown in (**b**), (**c**) and (**d**), respectively).

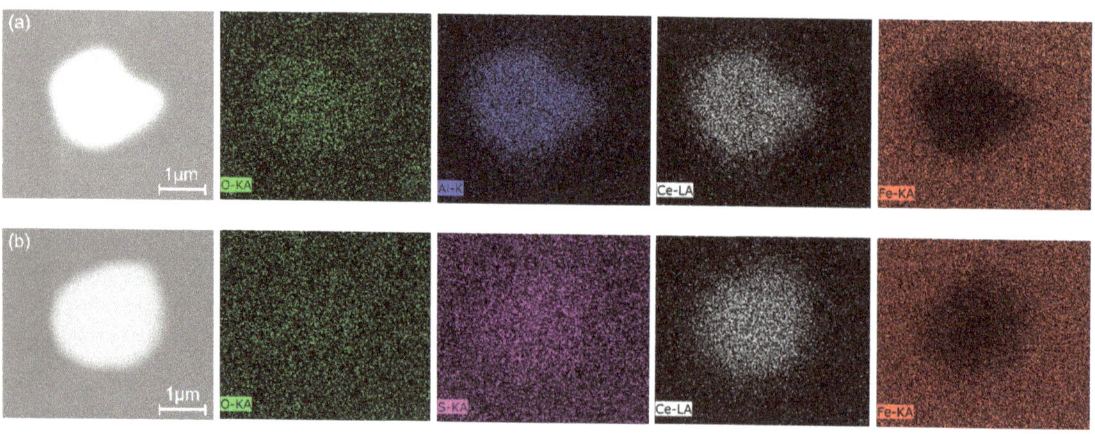

Figure 11. EDS element mappings of $CeAlO_3$ and Ce_2O_2S inclusions in the liquid metal pool during the ESR trial T2. (**a**) $CeAlO_3$ (**b**) Ce_2O_2S.

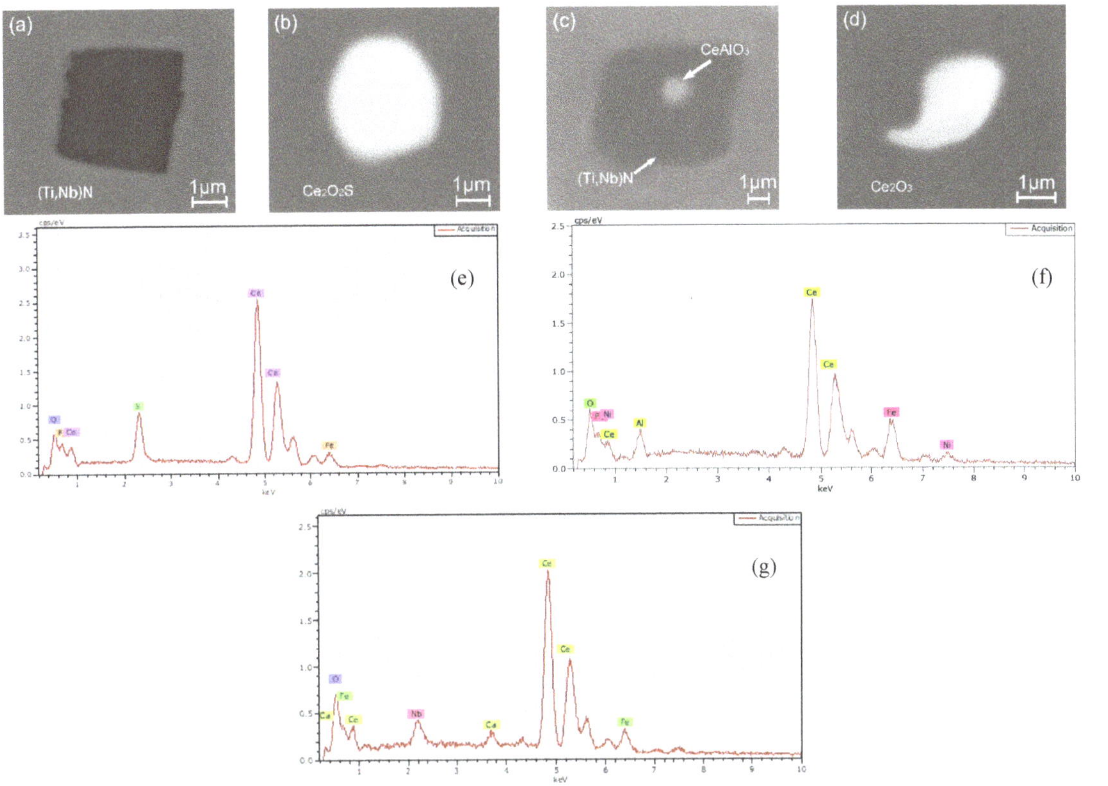

Figure 12. SEM images and EDS spectra of typical inclusions in the ingot C2. (**a**) (Ti,Nb)N, (**b**) Ce_2O_2S, (**c**) $CeAlO_3$ + (Ti,Nb)N, (**d**) Ce_2O_3. (EDS spectra shown in (**e**,**f**), and (**g**) correspond to the inclusions shown in (**b**,**c**), and (**d**), respectively).

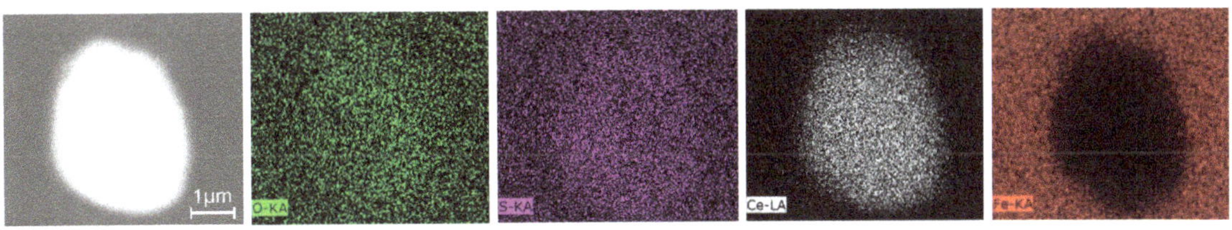

Figure 13. EDS element mappings of Ce_2O_2S inclusion in the ingot C2.

Typical inclusions in the liquid metal pool and remelted ingot for the ESR trial T3 are shown in Figures 14–17. The two types of rare earth inclusions in the liquid metal pool of the ESR trial T3 are observed to be Ce_2O_3 and Ce_2O_2S, and the morphology of these rare-earth inclusions is near-spherical. The results of the elements mappings of Ce_2O_2S inclusions in the liquid metal pool of ESR trial T3 and remelted ingot are shown in Figures 11 and 13, respectively.

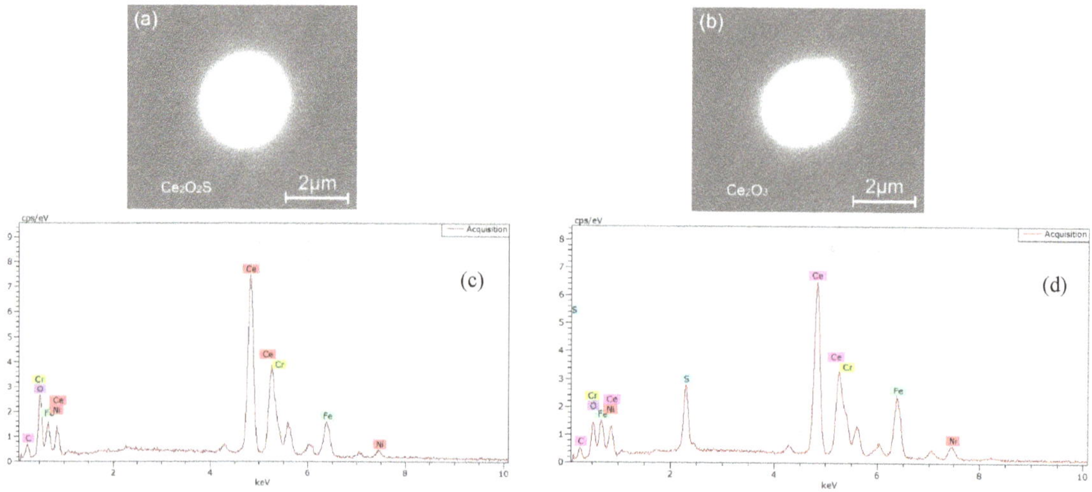

Figure 14. SEM images and EDS spectra of typical inclusions in the liquid metal pool during ESR trial T3. (**a**) Ce_2O_2S, (**b**) Ce_2O_3. (EDS spectra shown in (**c**) and (**d**)correspond to the inclusions shown in (**a**) and (**b**), respectively).

Figure 15. EDS element mappings of Ce_2O_2S inclusion in the liquid metal pool during ESR trial T3.

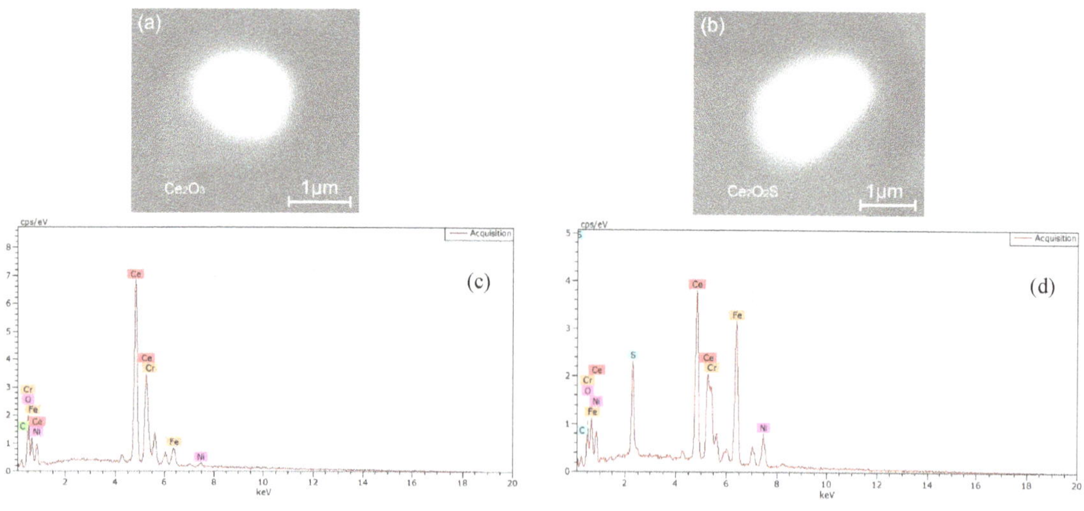

Figure 16. SEM images and EDS spectra of typical inclusions in the ingot C3. (**a**) Ce_2O_3, (**b**) Ce_2O_2S. (EDS spectra shown in (**c**) and (**d**) correspond to the inclusions shown in (**a**) and (**b**), respectively).

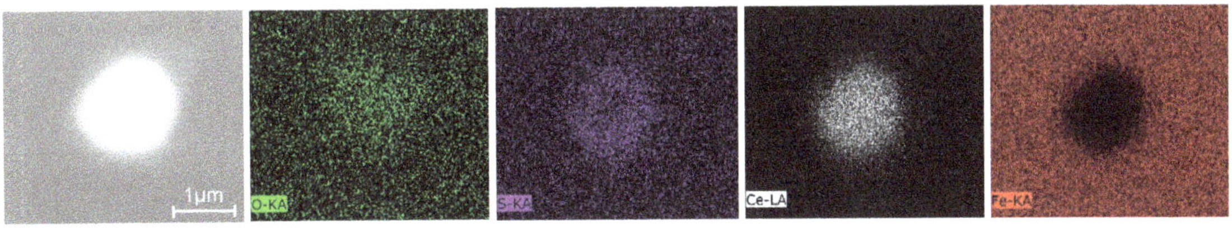

Figure 17. EDS element mappings of Ce_2O_2S inclusion in the ingot C3.

3.4. Size Distribution and Types of Inclusions

Most of inclusions in the Ce-free electrode are 2 to 5 μm in size (accounting for 72% in number proportion), followed by the inclusions larger than 5 μm (21% in the relative fraction), and the number proportion of the inclusion smaller than 2 μm is 7%. The size of inclusions in the electrode with 0.016 mass% Ce is 2 to 5 μm (accounting for 92% in number proportion). The number proportion of inclusions in the size range of 1 μm to 2 μm and the inclusions larger than 5 μm are 6% and 2%, respectively. The proportion of the inclusions in the size range of 1 μm to 2 μm in the electrode with 0.300 mass% Ce is 88%, and the number proportion of inclusions between 2 and 5 μm is 12%.

Figure 18 shows the size distribution of inclusions in liquid metal pools and remelted ingots with different Ce contents. The proportion of inclusions smaller than 2 μm in the liquid metal pool of ESR trial T1 is 67%. The proportion of the number of inclusions occupied between 2 μm and 5 μm is 31%. The size of inclusions smaller than 2 μm in ingot C1 decreased to 34%, and inclusions between 2 and 5 μm increased to 61%. The proportion of the number of inclusions in the size range of 1 μm to 2 μm in the liquid metal pool of the ESR trial T2 is 75%, and the proportion of the number of inclusions occupied between 2 and 5 μm is 25%. The proportion of inclusions in the size range of 1 μm to 2 μm in the remelted ingot decreased to 71%, and the proportion of inclusions in the size range of 2 to 5 μm increased to 27%. The proportion of inclusions larger than 5 μm is 2%. The proportion of inclusions in the liquid pool of the ESR trial T3 remained almost unchanged compared to the remelted ingot C3.

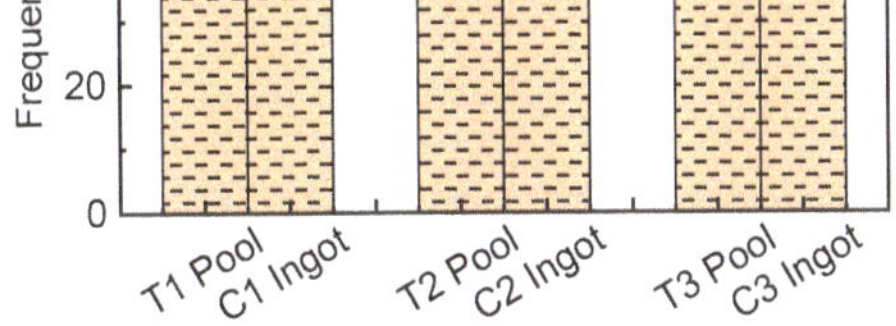

Figure 18. Size distribution of inclusions in the electrode, liquid metal pool, and ingots. (Pool represents liquid metal pool).

Figure 19 shows the distribution of inclusions types in electrodes, liquid metal pools, and remelted ingots with different Ce contents. All inclusions in the Ce-free electrode are identified as $MgO{\cdot}Al_2O_3$. The type of inclusions in the liquid metal pool of the ESR trial T1 has changed compared to the electrode. The newly formed $CaO–Al_2O_3$ inclusions in the liquid metal pool account for 6% in the number proportion. The proportion of $CaO–Al_2O_3$ inclusions in the ingot account for 49% of the total inclusions.

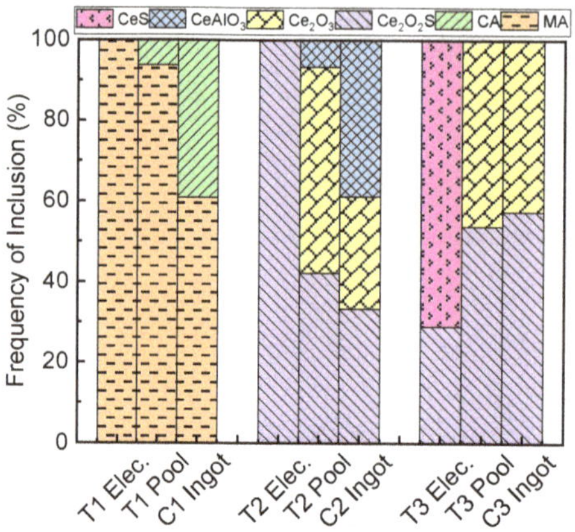

Figure 19. Proportion of different types of inclusion in steel (Elec. and Pool represent electrode and liquid metal pool, respectively. CA and MA represent CaO–Al_2O_3 and MgO·Al_2O_3, respectively).

The rare earth oxide inclusions in the electrode with 0.016 mass% Ce are identified as Ce_2O_2S. The rare earth oxide inclusions present in the liquid metal pool of ESR trial T2 are identified as Ce_2O_3, $CeAlO_3$, and Ce_2O_2S, accounting for 51%, 7%, and 42% in the number proportion, respectively. The Ce_2O_3 and $CeAlO_3$ inclusions are newly formed in the liquid metal pool. The proportion of Ce_2O_3 inclusions in the liquid metal pool and ingot reduce from 51% to 28% and $CeAlO_3$ inclusions increased from 7% to 39%. The Ce_2O_2S inclusions are reduced to 33% in the number proportion after ESR.

The CeS inclusions are completely removed and Ce_2O_3 inclusions are newly formed in the trial T3 during ESR. The proportion of Ce_2O_3 and Ce_2O_2S inclusions in the liquid metal pool is 46% and 54%, respectively. The relative proportions of inclusions in the remelted ingot (43% Ce_2O_3 and 57% Ce_2O_2S) do not change significantly compared to the metal pool.

3.5. Evolution Mechanism of Inclusions during Remelted Remelting

In the present study, thermodynamic calculation is performed for understanding the evolution of inclusions during the ESR with FactSage 8.0 (FSstel, FactPS, and FToxid database). Figures 20 and 21 show the transformation of inclusions precipitation with temperature for electrodes with different Ce content and remelted ingots C1, C2, and C3, respectively.

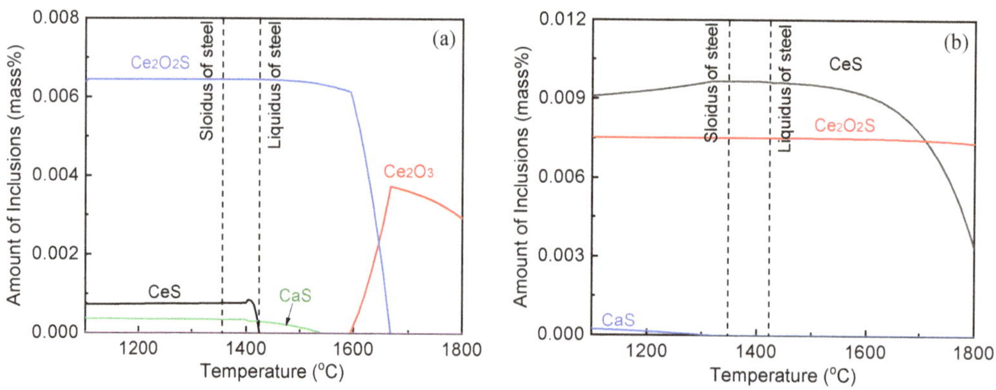

Figure 20. Inclusions formation in the electrode with different Ce contents calculated with FactSage 8.0. (**a**) 0.016 mass% Ce, (**b**) 0.300 mass% Ce.

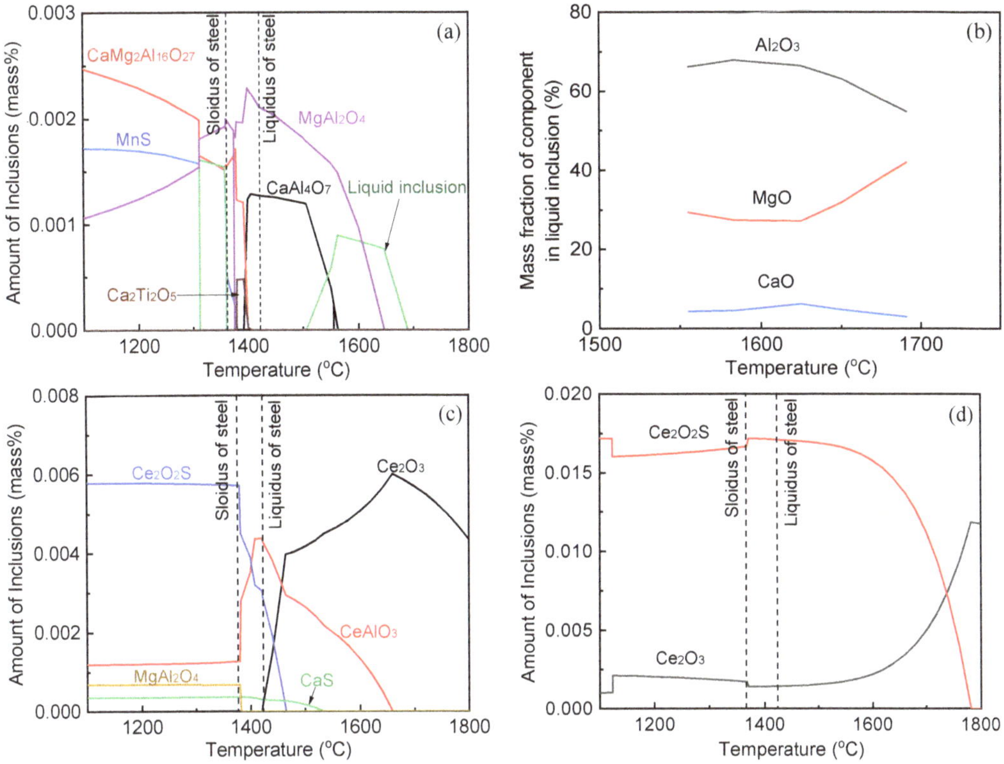

Figure 21. Inclusions transformation in the ingots with different Ce contents calculated with FactSage 8.0. (**a**) 0 mass% Ce, (**c**) 0.0055 mass% Ce, (**d**) 0.063 mass% Ce. (**b**) The change in the mass fraction of the component in liquid inclusion with 0 Ce.

According to the above experimental observations, no other types of oxide inclusions are observed in the Ce-free electrodes except for $MgAl_2O_4$ inclusions. The temperature of the liquid metal film steel at the electrode tip is close to the liquidus temperature during ESR [36,37]. The results showed that the part of $MgAl_2O_4$ inclusions are removed by the adsorption of the molten slag, and the remaining is not removed and entered into the liquid metal pool.

The predicted inclusions transformation with respect to the temperature in the electrodes is present in Figure 20. The Ce_2O_3 inclusions are already formed when the liquid steel temperature at 1800 °C. The Ce_2O_3 inclusions gradually precipitate out as the temperature decrease. when the temperature decreases to 1669 °C, the mass fraction of Ce_2O_3 inclusions decrease sharply, and the mass fraction of Ce_2O_2S inclusions increases sharply. The precipitation rate of Ce_2O_2S inclusions become slow when the temperature of the liquid steel decreases to 1593 °C. According to the thermodynamic calculation, the inclusion in the electrode containing 0.016 mass% Ce is mainly Ce_2O_2S, which is consistent with the experimental results.

It suggested that Ce_2O_2S inclusions dissociated into liquid steel as soluble oxygen, sulfur, and cerium as the liquid metal films formed on the downside of the electrode tip and collected into liquid metal droplets subsequently during ESR, whereas the others do not dissociate into liquid steel. Experimental results show that the Ce_2O_2S inclusions are not completely removed. The removal of these Ce_2O_2S inclusions is attributed to the dissociation of a portion of these original oxide inclusions into liquid steel and the absorption of the others by the molten slag phase during the ESR process.

The predicted inclusion transformation with respect to the temperature in the electrode with 0.300 mass% Ce is present in Figure 20b. The amount of the CeS inclusions is higher than that of Ce_2O_2S inclusions. The present thermodynamic calculation supports the above theoretical analysis of oxide inclusions formation and the experimental observations of oxide inclusions in the electrodes.

The electrode tip is heated to near the liquidus temperature of the steel during the ESR. As shown in Figure 20b, CeS and Ce_2O_2S inclusions in the electrode with 0.300 mass% Ce do not dissociate into liquid steel as soluble elements. The experimental results show that CeS inclusions are fully removed by the adsorption of the molten slag before liquid metal droplets collect in the liquid metal pool. According to the results of the experiment, a portion of the Ce_2O_2S inclusions are removed and the remaining enters into the liquid metal pool. The removal of these Ce_2O_2S inclusions is attributed to the absorption of the molten slag during the ESR process.

The predicted inclusions transformation with respect to the temperature in the remelted ingot C1 is present in Figure 21a. The liquid inclusions start to precipitate in the liquid steel at 1690 °C. As the temperature decreases, the mass fraction of liquid inclusions increases sharply. The precipitation rate of liquid inclusions slows down at which point $MgAl_2O_4$ inclusions begin to precipitate when the temperature is reduced to 1647 °C. When the temperature is reduced to 1574 °C, the mass fraction of liquid inclusions decreases and $CaAl_4O_7$ inclusions start to precipitate. Figure 21b shows the variation of the mass fraction of each component in the liquid inclusions with temperature. It can be seen from Figure 21b that the liquid inclusions consist of Al_2O_3, MgO, and CaO, in which Al_2O_3 is dominant in the mass fraction.

As shown in Tables 1 and 2, the reoxidation of liquid steel occurs during the ESR process of trial T1. Thereafter, soluble oxygen reacts with calcium, aluminum and magnesium to form $CaAl_4O_7$ and $MgAl_2O_4$ inclusions in the liquid metal pool during the ESR.

All oxide inclusions in the Ce-free electrode were identified as $MgAl_2O_4$. The types of inclusions in the liquid metal pool after ESR are $MgAl_2O_4$ (accounting for 94% in number proportion) and $CaAl_4O_7$ (6% in number proportion). The proportion of these two types of inclusions in the remelted ingot C1 is 61% and 39%, respectively. Thermodynamic calculations support the above analysis of oxide inclusions in the remelted ingot C1.

The predicted inclusions transformation with respect to the temperature in the ingot C2 is present in Figure 21c. The Ce_2O_3 inclusions already form at 1800 °C in the liquid steel, and Ce_2O_3 inclusions precipitate slowly as the temperature gradually decreases. The mass fraction of Ce_2O_3 inclusions decreases rapidly at 1659 °C, and $CeAlO_3$ inclusions start to form in the liquid steel.

As listed in Tables 1 and 2, the changes in oxygen content before and after the ESR show that reoxidation of liquid steel occurred in the trial T2 during the ESR. The soluble oxygen inside the liquid metal pool during trial T2 ESR steel reacts with cerium $2[Ce] + 3[O] = Ce_2O_3(s)$ ($\Delta G^o = -1431120 + 121T$ (J/mol)) [38] to form Ce_2O_3 and the Gibbs free energy change of the reaction to -14.56 kJ/mol. Once the soluble oxygen pickup takes place in liquid metal pool (arising from the reoxidation), the soluble oxygen reacts with soluble aluminum and cerium in liquid metal pool to form $CeAlO_3$ inclusions, $[Ce] + [Al] + 3[O] = CeAlO_3(s)$ ($\Delta G^o = -1366460 + 364T$ (J/mol)) [39] with a Gibbs free energy change of -41.47 kJ/mol. The present thermodynamic calculation supports the above analysis of oxide inclusions formation and the experimental observations of oxide inclusions in the liquid metal pool and ingots.

The predicted inclusions transformation with respect to the temperature in ingot C3 with respect to the temperature is present in Figure 21d. The Ce_2O_3 inclusions are already formed at 1800 °C. When the temperature is reduced to 1781 °C, the mass fraction of Ce_2O_3 inclusions starts to decrease in the liquid steel and the Ce_2O_2S inclusions start to precipitate in the liquid steel.

The change of oxygen content in Tables 1 and 2 shows that reoxidation of liquid steel occurs during ESR. The reoxidation of liquid steel leads to considerable pickup of soluble oxygen during ESR, which provides a driving force for the generation of Ce_2O_3 inclusions. $2[Ce] + 3[O] = Ce_2O_3(s)$ ($\Delta G^o = -1431120 + 121T$ (J/mol)), the Gibbs free energy change is −142.37 kJ/mol. Soluble oxygen in the liquid metal pool reacts with soluble sulfur and cerium $2[Ce] + 2[O] + [S] = Ce_2O_2S(s)$ ($\Delta G^o = -1353590 + 332T$ (J/mol)) [40] to form Ce_2O_2S inclusions. The Gibbs free energy change is −158.85 kJ/mol. According to the experimental results, the Ce_2O_2S inclusions in the electrode cannot be removed completely, the Ce_2O_2S inclusions in the liquid metal pool originate from reoxidation inside the liquid metal pool and the original inclusions that has not been removed. The thermodynamic calculations results are consistent with the experimental observations and analysis of the formation of oxide inclusions in the liquid metal pool and remelted ingot.

The liquid steel solidification progresses in a is non-equilibrium state during the ESR practice, which leads to insufficient solute diffusion in solid [41,42]. The evolution of inclusions during the solidification of liquid steel with different Ce contents is calculated using the non-equilibrium module (Scheil-Gulliver model) of the thermodynamic software FactSage 8.0.

As shown in Figure 22a, $MgAl_2O_4$ and $CaAl_4O_7$ inclusions are formed at the high-temperature stage (around 1600 °C). The mass fraction of oxide inclusions forms at temperatures below the liquidus temperature of the steel is virtually nothing during the solidification process. Figure 22b shows that the Ce_2O_3 inclusions already formed at 1600 °C in the liquid steel. The mass fraction of Ce_2O_3 inclusions no longer changes as the temperature decreases. When the temperature reaches to 1600 °C, $CeAlO_3$ inclusions are formed in liquid steel and the mass fraction of oxide inclusions does not change when the temperature is reduced to 1449 °C. At the temperature below the liquidus temperature of the steel, the mass fraction of oxide inclusions does not change. As shown in Figure 22c, the Ce_2O_3 and Ce_2O_2S inclusions are formed at 1600 °C in liquid steel. As the temperature of liquid steel decreases, the mass fraction of Ce_2O_3 and Ce_2O_2S inclusions no longer change.

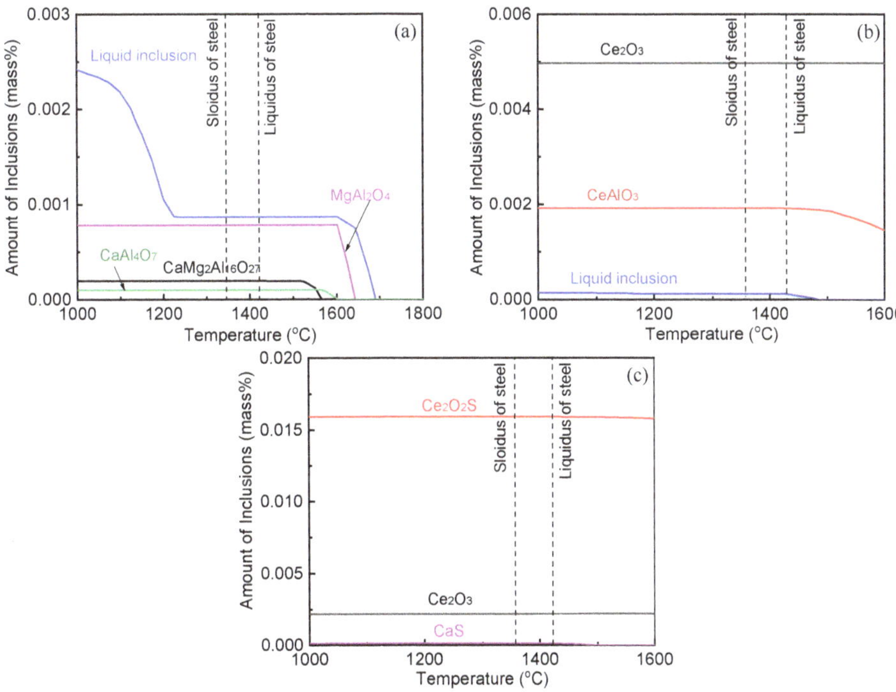

Figure 22. (**a**) Non-equilibrium precipitation of inclusions during the cooling and solidification of liquid steel. (**a**) 0 mass% Ce, (**b**) 0.0055 mass% Ce, (**c**) 0.063 mass% Ce.

As shown in Figure 22, amount of individual type of oxide inclusions does not change during liquid steel solidification. It can be learned that no fresh oxide inclusions are generated during the solidification of liquid steel. The types of inclusions in the liquid metal pools of the ESR trials T1, T2, and T3 are the same as those in the remelted ingots. This is consistent with the experimental observations. The differences in the number proportions of different types of inclusions between liquid metal pool and remelted ingot are attributed to the removal through floatation before full solidification of liquid steel during the ESR.

4. Conclusions

1. The Al pickup in the steel is caused by the Al addition for deoxidation during the ESR process, rather than the reduction of Al_2O_3 in the slag by Ce. The soluble oxygen pickup is generated in liquid steel due to the decomposition of FeO in the slag and desulfurization during the protective argon gas atmosphere ESR.
2. The oxide inclusions in Ce-free electrode are $MgO\cdot Al_2O_3$, part of which are removed by molten slag absorption during the ESR. The oxide inclusions in liquid metal pool are mainly $MgO\cdot Al_2O_3$ and $CaO–Al_2O_3$ (6% in number fraction). The soluble oxygen that arising from reoxidation of liquid steel during the ESR react with soluble calcium and aluminum to form $CaO–Al_2O_3$ inclusions. $MgO\cdot Al_2O_3$ inclusions are originated from reoxidation products and the relics from the electrode.
3. The oxide inclusions in the electrode with 0.016 mass% Ce are Ce_2O_2S. Part of Ce_2O_2S inclusions are removed during ESR in two ways: (I) dissociated into soluble oxygen and soluble elements in liquid steel, (II) absorbed by molten slag. The oxide inclusions in the liquid metal pool are Ce_2O_3, $CeAlO_3$, and Ce_2O_2S. $CeAlO_3$ inclusions are reoxidation product, and Ce_2O_2S inclusions are the relics from the electrode. The proportions of Ce_2O_3, $CeAlO_3$ and Ce_2O_2S inclusions in ingot are 28%, 39%, and 33%, respectively.
4. The rare-earth inclusions in the electrode with 0.300 mass% Ce are Ce_2O_2S and CeS. The CeS inclusions are fully removed during ESR. Part of Ce_2O_2S inclusions are removed by slag adsorption. The oxide inclusions in liquid metal pool are Ce_2O_3 (reoxidation products, 46% in number proportion) and Ce_2O_2S (54% in number fraction). Ce_2O_2S inclusions in liquid metal pool are originated from the relics of electrode and reoxidation products. The proportion of Ce_2O_3 and Ce_2O_2S inclusions in remelted ingots are 43% and 57%, respectively.
5. No fresh oxide inclusions are generated during the solidification of liquid steel. The differences in the number proportions of different types of inclusions between liquid metal pool and remelted ingot are attributed to the removal through floatation before full solidification of liquid steel during the ESR.

Author Contributions: Conceptualization, C.S.; methodology, Z.W. and X.Z.; formal analysis, Z.W. and C.S.; investigation, Z.W. and C.S.; writing—original draft preparation, Z.W., S.W. and C.S.; writing—review and editing, C.S. and X.Z.; supervision, C.S. and J.L.; funding acquisition, C.S. All authors have read and agreed to the published version of the manuscript.

Funding: This research was funded by the National Natural Science Foundation of China (Grant Nos. 52074027 and 51874026).

Institutional Review Board Statement: Not applicable.

Informed Consent Statement: Not applicable.

Data Availability Statement: Not applicable.

Acknowledgments: The authors would like to express appreciation to the National Natural Science Foundation of China for the financial support.

Conflicts of Interest: The authors declare no conflict of interest.

References

1. Kondrat'Ev, S.Y.; Kraposhin, V.S.; Anastasiadi, G.P.; Talis, A.L. Experimental observation and crystallographic description of M_7C_3 carbide transformation in Fe–Cr–Ni–C HP type alloy. *Acta Mater.* **2015**, *100*, 275–281. [CrossRef]
2. Hedaiat, F.; Dehmolaei, R.; Khorasanian, M.; Lotfi, B. Long-term oxidation behaviour and thermal stability of heat-resistant stainless steel claddings deposited on AISI 316 stainless steel by the GTAW process. *Surf. Coat. Technol.* **2021**, *424*, 127605. [CrossRef]
3. Mazánová, V.; Heczko, M.; Polák, J. Fatigue crack initiation and growth in 43Fe-25Ni-22.5Cr austenitic steel at a temperature of 700 °C. *Int. J. Fatigue* **2018**, *114*, 11–21. [CrossRef]
4. Guo, H.; Yang, S.F.; Wang, T.T.; Yuan, H.; Zhang, Y.L.; Li, J.S. Microstructure evolution and acicular ferrite nucleation in inclusion-engineered steel with modified MgO@C nanoparticle addition. *J. Mater. Sci. Technol.* **2022**, *99*, 277–287. [CrossRef]
5. Xu, X.; Benaarbia, A.; Allen, D.J.; Jepson, M.A.E.; Sun, W. Investigation of microstructural evolution and creep rupture behaviour of 9% Cr MarBN steel welds. *Mater. Sci. Eng. A* **2020**, *791*, 139546. [CrossRef]
6. Yang, Z.; Zhang, J.; Li, S.; Li, G.; Wang, Q.; Hui, W.; Weng, Y. On the critical inclusion size of high strength steels under ultra-high cycle fatigue. *Mater. Sci. Eng. A* **2006**, *427*, 167–174. [CrossRef]
7. Meurling, F.; Melander, A.; Tidesten, M.; Westin, L. Influence of carbide and inclusion contents on the fatigue properties of high speed steels and tool steels. *Int. J. Fatigue* **2001**, *23*, 215–224. [CrossRef]
8. Shibaeva, T.V.; Laurinavichyute, V.K.; Tsirlina, G.A.; Arsenkin, A.M.; Grigorovich, K.V. The effect of microstructure and non-metallic inclusions on corrosion behavior of low carbon steel in chloride containing solutions. *Corros. Sci.* **2014**, *80*, 299–308. [CrossRef]
9. Tomita, Y. Effect of morphology of nonmetallic inclusions on tensile properties of quenched and tempered 0.4C-Cr-Mo-Ni steel. *Mater. Charact.* **1995**, *34*, 121–128. [CrossRef]
10. Garrison, W.M.; Wojcieszynski, A.L. A discussion of the effect of inclusion volume fraction on the toughness of steel. *Mater. Sci. Eng. A.* **2007**, *464*, 321–329. [CrossRef]
11. Shi, C.B.; Wang, S.J.; Li, J.; Cho, J.W. Non-metallic inclusions in electroslag remelting: A review. *J. Iron Steel. Res. Int.* **2021**, *28*, 1483–1503. [CrossRef]
12. Wang, M.J.; Mu, S.M.; Sun, F.F.; Wang, Y. Influence of rare earth elements on microstructure and mechanical properties of cast high-speed steel rolls. *J. Rare Earths* **2007**, *25*, 490–494. [CrossRef]
13. Chen, R.C.; Wang, Z.G.; He, J.G.; Zhu, F.S.; Li, C.H. Effects of rare earth elements on microstructure and mechanical properties of H13 die steel. *Metals* **2020**, *10*, 918. [CrossRef]
14. Wang, L.M.; Du, X.J.; Gan, Y.; Liu, L.; Ye, X.N.; Jiang, L.Z. Effect of rare earths on oxidation resistance of heat resistant steel. *J. Rare Earths* **2010**, *28*, 489–491. [CrossRef]
15. Gao, J.Z.; Fu, P.X.; Liu, H.W.; Li, D.Z. Effects of rare earth on the microstructure and impact toughness of H13 Steel. *Metals* **2015**, *5*, 383–394. [CrossRef]
16. Zhang, S.C.; Yu, J.T.; Li, H.B.; Jiang, Z.H.; Geng, Y.F.; Feng, H.; Zhang, B.B.; Zhu, H.C. Refinement mechanism of cerium addition on solidification structure and sigma phase of super austenitic stainless steel S32654. *J. Mater. Sci. Technol.* **2022**, *102*, 105–114. [CrossRef]
17. Xu, Y.W.; Song, S.; Wang, J.W. Effect of rare earth cerium on the creep properties of modified 9Cr–1Mo heat-resistant steel. *Mater. Lett.* **2015**, *161*, 616–619. [CrossRef]
18. Wang, H.P.; Xiong, L.; Zhang, L.; Wang, Y.; Shu, Y.Y.; Zhou, Y. Investigation of RE-O-S-As inclusions in high carbon steels. *Metall. Mater. Trans. B* **2017**, *48*, 2849–2858. [CrossRef]
19. Liu, Y.Y.; Bao, X.R.; Chen, L.; Jin, Z.L.; Ren, H.P. Rare Earth microalloied elements′ influence on the organization and capability of X65 pipeline steel. *J. Rare Earths* **2010**, *28*, 497–500. [CrossRef]
20. Chen, L.; Ma, X.C.; Wang, L.M.; Ye, X.N. Effect of rare earth element yttrium addition on microstructures and properties of a 21Cr–11Ni austenitic heat-resistant stainless steel. *Mater. Des.* **2011**, *32*, 2206–2212. [CrossRef]
21. Zhao, Y.Y.; Wang, J.F.; Zhou, S.; Wang, X.D. Effects of rare earth addition on microstructure and mechanical properties of a Fe–15Mn–1.5Al–0.6C TWIP steel. *Mater. Sci. Eng. A* **2014**, *608*, 106–113. [CrossRef]
22. Wang, Y.G.; Liu, C.J. Agglomeration characteristics of various inclusions in Al-killed molten steel containing rare earth element. *Metall. Mater. Trans. B* **2020**, *51*, 2585–2595. [CrossRef]
23. Huang, Y.; Cheng, G.G.; Li, S.J.; Dai, W.X. Effect of cerium on the behavior of inclusions in H13 steel. *Steel Res. Int.* **2018**, *89*, 1800371. [CrossRef]
24. Liu, H.H.; Fu, P.X.; Liu, H.W.; Cao, Y.F.; Sun, C.; Du, N.Y.; Li, D.Z. Effects of rare earth elements on microstructure evolution and mechanical properties of 718H pre-hardened mold steel. *J. Mater. Sci. Technol.* **2020**, *50*, 245–256. [CrossRef]
25. Feng, H.; Lu, P.C.; Li, H.B.; Jiang, Z.H. Effect of Mg pretreatment and Ce addition on cleanliness and inclusion evolution in high-nitrogen stainless bearing steels. *Metall. Mater. Trans. B* **2022**, *53*, 864–876. [CrossRef]
26. Geng, R.M.; Li, J.; Shi, C.B. Effect of Ce on inclusion evolution and HAZ mechanical properties of Al-killed high-strength steel. *Ironmak. Steelmak.* **2021**, *48*, 796–802. [CrossRef]
27. Geng, R.M.; Li, J.; Shi, C.B. Evolution of inclusions with Ce addition and Ca treatment in Al-killed steel during RH refining process. *ISIJ Int.* **2021**, *61*, 1506–1513. [CrossRef]

28. Shi, C.B.; Yu, W.T.; Wang, H.; Li, J.; Jiang, M. Simultaneous modification of alumina and $MgO{\cdot}Al_2O_3$ inclusions by calcium treatment during electroslag remelting of stainless tool steel. *Metall. Mater. Trans. B* **2016**, *48*, 146–161. [CrossRef]
29. Li, X.; Jiang, Z.H.; Geng, X.; Chen, M.J.; Cui, S. Effect of rare earth–magnesium alloy on inclusion evolution in industrial production of die steel. *Steel Res. Int.* **2019**, *90*, 1900103. [CrossRef]
30. Shi, C.B.; Zheng, D.L.; Guo, B.S.; LI, J.; Jiang, F. Evolution of oxide–sulfide complex inclusions and its correlation with steel cleanliness during electroslag rapid remelting (ESRR) of tool steel. *Metall. Mater. Trans. B* **2018**, *49*, 3390–3402. [CrossRef]
31. Wang, S.J.; Shi, C.B.; Liang, Y.J.; Wan, X.X.; Zhu, X. Evolution and formation of non-metallic inclusions during electroslag remelting of a heat-resistant steel for ultra-supercritical power plants. *Metall. Mater. Trans. B* **2022**, *53*, 3095–3114. [CrossRef]
32. Gao, S.; Wang, M.; Guo, J.L.; Wang, H.; Zhi, J.G.; Bao, Y.P. Characterization transformation of inclusions using rare earth Ce treatment on Al-Killed titanium alloyed interstitial free steel. *Steel Res. Int.* **2019**, *90*, 1900194. [CrossRef]
33. Geng, R.M.; Li, J.; Shi, C.B.; Zhi, J.G.; Lu, B. Effect of Ce-La on inclusion evolution in Al-killed high strength steel. *Metall. Res. Technol.* **2020**, *117*, 616. [CrossRef]
34. Shi, C.B.; Wang, H.; Li, J. Effects of reoxidation of liquid steel and slag composition on the chemistry evolution of inclusions during electroslag remelting. *Metall. Mater. Trans. B* **2018**, *49*, 1675–1689. [CrossRef]
35. Shi, C.B. Deoxidation of electroslag remelting (ESR)—A review. *ISIJ Int.* **2020**, *60*, 1083–1096. [CrossRef]
36. Liu, H.L.; Liu, C.J.; Jiang, M.F. Effect of rare earths on impact toughness of a low-carbon steel. *Mater. Des.* **2012**, *33*, 306–312. [CrossRef]
37. Kruger, D.; Garbers, C.A. Characteristics and modification of non-metallic inclusions in titanium-stabilized aisi 409 ferritic stainless steel. *Metall. Mater. Trans. B* **2017**, *48*, 1514–1532. [CrossRef]
38. Wen, T.J.; Ren, Q.; Zhang, L.F.; Wang, J.J.; Ren, Y.; Zhang, J.; Yang, W.; Xu, A.J. Evolution of nonmetallic inclusions during the electroslag remelting process. *Steel Res. Int.* **2021**, *92*, 2000629. [CrossRef]
39. Ren, Q.; Zhang, L.F. Effect of cerium content on inclusions in an ultra-low-carbon aluminum-killed steel. *Metall. Mater. Trans. B* **2020**, *51*, 589–600. [CrossRef]
40. Adabavazeh, Z.; Hwang, W.S.; Su, Y.H. Effect of adding cerium on microstructure and morphology of Ce-based inclusions formed in low-carbon steel. *Sci. Rep.* **2017**, *7*, 46503. [CrossRef]
41. Shi, C.B.; Zheng, X.; Yang, Z.B.; Lan, P.; Li, J.; Jiang, F. Effect of melting rate of electroslag rapid remelting (ESRR) on the microstructure and carbides in a hot work tool steel. *Metals Mater. Int.* **2021**, *27*, 3603–3616. [CrossRef]
42. Klancnik, U.; Kosec, B.; Mrvar, P.; Medved, J. Non-equilibrium solidification and microsegregation in centrifugally cast high speed steel for rolls. *J. Min. Metall. B* **2018**, *54*, 59–66. [CrossRef]

metals

MDPI

Article

Effect of Sulfur Content on the Inclusion and Mechanical Properties in Ce-Mg Treated Resulfurized SCr420H Steel

Meng Sun [1], Zhouhua Jiang [1,2,*], Yang Li [1,*], Changyong Chen [3], Shuai Ma [1], Yongshuai Ji [1], Ju Wang [1] and Hang Liu [4]

1 School of Metallurgy, Northeastern University, Shenyang 110819, China; 1810517@stu.neu.edu.cn (M.S.); mashuai2408@gmail.com (S.M.); 2001535@stu.neu.edu.cn (Y.J.); 2071604@stu.neu.edu.cn (J.W.)
2 State Key Laboratory of Rolling and Automation, Northeastern University, Shenyang 110819, China
3 School of Materials and Metallurgy, Wuhan University of Science and Technology, Wuhan 430081, China; cychen@wust.edu.cn
4 Institute of Metal Research, Chinese Academy of Sciences, Shenyang 110016, China; liuhang@imr.ac.cn
* Correspondence: jiangzh@smm.neu.edu.cn (Z.J.); liy@smm.neu.edu.cn (Y.L.)

Abstract: To clarify the effect of sulfur on inclusions and mechanical properties of Ce-Mg treated resulfurized SCr420H steel. Laboratory experiments were conducted to prepare steels with sulfur contents as 0.01%, 0.06%, and 0.132%. Inclusion evolution in liquid steel, MnS precipitation during solidification, and tensile test results of steel after quenching and tempering were investigated. The results showed that due to the limitation of mass transfer in molten steel, composite inclusion that Ce-O-S wrapped by Ce-Ca-Mg-Al-Si-O, which was named transition state inclusions, can form quickly after adding Ce-Mg lump to the molten steel. As the homogenization of molten steel, the difference of sulfur content in steel can lead to the transition state inclusions transformed into different inclusions. With the increase of sulfur content, the quantity of MnS increased significantly, and the morphology of MnS transformed from "stick" to "dendritic + fishbone", and then to "fishbone". Tensile test results and fracture analysis indicate that the decline of inclusion spacing as the increase of sulfur content leads to a shorter physical path of crack propagation in steel. Therefore, the increase of sulfur content can bring about a decrease in the strength and plasticity of the steel. From the perspective of inclusion control, making the MnS inclusion precipitate more dispersive and increasing the distance between inclusions can be considered as a method for preventing the decline of mechanical properties in steel with high sulfur content.

Keywords: MnS; inclusion; cerium; sulfur; fracture

Citation: Sun, M.; Jiang, Z.; Li, Y.; Chen, C.; Ma, S.; Ji, Y.; Wang, J.; Liu, H. Effect of Sulfur Content on the Inclusion and Mechanical Properties in Ce-Mg Treated Resulfurized SCr420H Steel. *Metals* **2022**, *12*, 136. https://doi.org/10.3390/met12010136

Academic Editor: Yanling Zhang

Received: 23 November 2021
Accepted: 7 January 2022
Published: 11 January 2022

1. Introduction

As an element that can improve steel machinability, sulfur has been widely used in gear steel, non-quenched and tempered steel, and other steel grades that require good turning and forming performance [1,2]. The increase of sulfur content can result in the formation of a large amount of MnS in steel. This inclusion can play a lubricating role during machining, reducing tool wear and machining costs [3–5]. Resulfurized gear steel SCr420H produced by a steel plant in China is a steel grade with sulfur addition (0.01–0.02%), was widely used in the manufacture of high-grade automobile gears. The production process was as follows: electric arc furnace (EAF)→ladle furnace (LF)→vacuum degassing furnace (VD)→continuous casting (CC). During the production process, sulfur addition was conducted at the final timing of the VD process. Due to the lack of machinability, it is considered to design an enhanced machinability gear steel based on steel SCr420H with increasing sulfur addition. In addition, to achieve grain refinement during the gear carburizing process [6,7], grain refining elements cerium and magnesium are considered to be added into steel during the secondary refining process [8–16]. However, it is known that sulfur, as one of the common inclusion forming elements, can participate in inclusion formation in liquid steel [17,18] and MnS precipitation during solidification [19] and present an impact on the mechanical properties of steel finally [20]. Therefore, an in-depth understanding of the influence of sulfur content on inclusions and mechanical properties becomes necessary for designing a proper sulfur content for industrial production.

In recent years, the effect of cerium and magnesium in steel has gradually become a research hotspot, and many studies have reported that cerium and magnesium can improve the cleanliness and properties of steel [21–27]. However, the application of these two elements in industrial production has some limitations. High cerium content can easily cause nozzle clogging [28], and magnesium treatment have problems such as low yield and severe reaction [29]. Under this background, Ce-Mg combined treatment technology has attracted the interest of scholars. This technology's idea is to realize cerium and magnesium's synergistic effect and avoid a high single element content addition. At present, this technology has been applied in the industrial processing of die steel for improving its performance [30]. As the elements with strong reactivity, cerium and magnesium can obviously impact inclusion in steel [31–37]. Li et al. [35] studied that the effect of Ce-Mg alloy on inclusion evolution in industrial production of die steel, and they found that with the increase of magnesium content, the stability area of cerium inclusions moves from the Ce_2O_2S stability area to the direction of the CeS stability area. The literature [36] indicates that rare earth magnesium alloy has better deoxidization and desulfurization ability than only adding cerium and only adding Mg in H13 steel, because the composite inclusions that low-density MgO attaching or wrapping on the surface of high-density cerium inclusions are easy to float up. Chang et al. [37] found that inclusions in GCr15 bearing steel were obviously refined after rare earth (cerium and lanthanum) and magnesium treatment, and the proportion of large particle inclusions decreases with increasing rare earth content. It is noted that some studies clarified the relations between the content of rare earth element and magnesium and inclusion evolution, but it is rarely seen that research about the role of sulfur content in inclusion formation and evolution after Ce-Mg treated steel.

As mentioned, MnS inclusion can enhance machinability as the free-cutting phase in steel. However, the mechanical properties of steel can inevitably be damaged, which is contradictory to the requirement of gear steel to have enough strength and plasticity. Many studies have studied the harmful effect of MnS inclusion in steel, but these researches mainly focus on steel with low sulfur content [38–41]. In resulfurized steel, the relation between MnS precipitation and fracture in steel with high sulfur content is needed to research, especially it may be different from steel with low sulfur content, due to the variety of morphology, quantity, and distribution of MnS.

In the present study, the industrial production process of resulfurized SCr420H was simulated in laboratory. Three groups of steels with different sulfur content were prepared for comparison. Cerium and magnesium were added to the liquid steel during the refining process, and the influence of sulfur content on inclusion evolution was analyzed by sampling. The three-dimensional morphologies of MnS inclusion were revealed by the method of electrolytic etching. Based on the tensile test results and fracture observation, the relation between MnS inclusion and steel fracture was proposed.

2. Methods

2.1. Material Preparation

In this experiment, cerium and magnesium powder were mixed evenly and then pressed into a lump by using a powder compressing machine. The metal ratio was (5% Ce-20% Mg-75% Fe), and the addition amount in the experiment was set as 0.7%. Chemical reagents (Sinopharm Chemical Reagent Co., Ltd., Shanghai, China) were used to prepare refining slag, and all reagents were heated at 800 °C for 6 h to remove moisture. To simulate the actual production situation, graphite, ferrosilicon, ferromanganese, chromium, nickel, copper, and molybdenum supplied by a domestic steel plant were used as alloy materials in this experiment, and MgO crucible was used as the reaction vessel (Shandong Refractories

Group Lunai Kiln Refractory Co., Ltd., Shandong, China). The MgO crucible has a height of 90 mm and a diameter of 55 mm. The total mass of experimental raw materials was controlled at about 1000 g.

2.2. Experiment Procedure

Three experimental steels with different sulfur content were designed, each named Steel A, Steel B, and Steel C. Sulfur was added in steel in the form of FeS, and the addition of sulfur in the experiment was designed as 150 ppm, 700 ppm, and 1500 ppm, respectively. The experiment was conducted in a $MoSi_2$ tube furnace, as shown in Figure 1. To control the air tightness of furnace, a simple air tightness inspection device was installed at the outlet of the furnace. During the whole experiment, 99.999% of high purity argon was continuously introduced into the furnace to control the protective atmosphere. The MgO crucible was placed in a graphite crucible, and the role of graphite crucible was to protect MgO crucible. The flow chart of the experiment was shown in Figure 2. The total time of the experiment was controlled at 73 min. Before the experiment began, the molybdenum rod was used to stir the molten steel several times to ensure that the molten steel was completely liquid and uniform. At 5 min, Al was fixed in the bottom of quartz tube and inserted in the molten steel during experiments. Refining slag (55% CaO 11% SiO_2 29% Al_2O_3 5% MgO Basicity R = 5) was added at 13 min of the experiment. At 48 min, the refining process enters the second stage. Slag with high SiO_2 content was added on steel for the purpose of adjusting the slag basicity to 1.5. At 53 min, FeS wrapped by iron foil was fixed in the bottom of quartz tube and inserted in the molten steel. In order to ensure the yield, Ce-Mg lump is usually added at about 5–15 min before the end of steelmaking in production, and this design was referred to in the current experiment. Ce-Mg lump was added in liquid steel after FeS addition for 5 min. At 73 min, experiments were finished and started cooling. To understand the evolution behavior of inclusions in an experiment, quartz tubes with an internal diameter of 6 mm were used to extract the sample. Sample S0 was obtained at 58 min (timing before Ce-Mg lump addition), sample S1 was obtained at 60 min (timing after Ce-Mg lump addition for 2 min), and sample S2 was cut from the as-cast ingot. In order to ensure the validity of experimental data, the first group experiment (Steel A) was repeated.

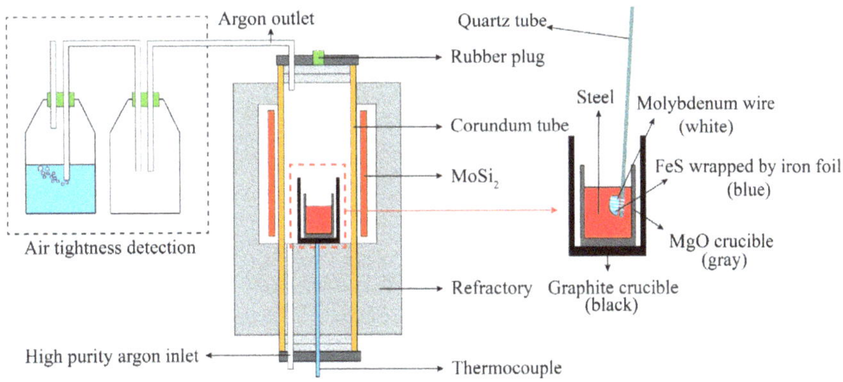

Figure 1. Sketch of experimental setup.

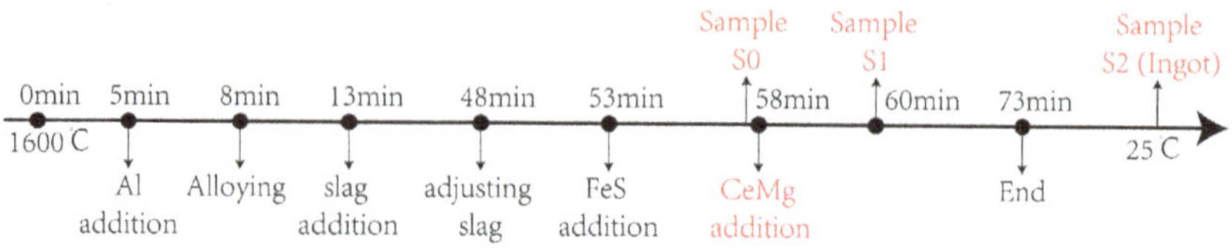

Figure 2. Experimental process.

The sample treatment method for inclusion observation and mechanical properties was shown in Figure 3. The cylindrical ingot was cut into two semi-cylinders along diameter. One was forged and conducted heat treatment for tensile test, and another was used to acquire samples for inclusion observation. The forging process starts at 1200 °C after heat preservation for 2 h, and finally air-cooled to room temperature after forging. The cross section of forging bar was 12 mm × 12 mm. Forged steel were austenized at 900 °C for 25 min, and oil quenched to room temperature, followed by tempering at 200 °C for 120 min. The atmosphere of heat treatment process was air. For the mechanical property characterization, tensile tests were conducted on Shimadzu AGS-X100KN electronic tensile testing machine following standard GB/T 228.1-2010 (ISO 6892-1:2009, MOD).

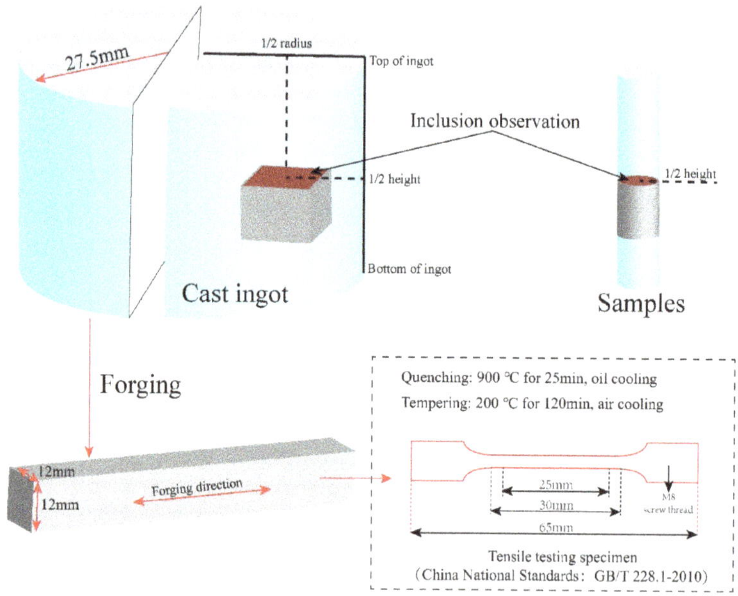

Figure 3. Sample treatment method for inclusion observation and mechanical properties.

2.3. Analysis

The as-cast sample was used to measure the steel composition. The contents of silicon, manganese, molybdenum, chromium, nickel and copper were measured by a direct reading spectrometer (ARL4460, Thermo Fisher Scientific, Waltham, MA, USA). The contents of carbon and sulfur were measured by infrared carbon and sulfur analyzer (CS230, LECO Corporation, St. Joseph, MI, USA), the contents of cerium and acid solution aluminum were measured by ICP-OES (Optima 8300DV, PerkinElmer, Waltham, MA, USA), The content of magnesium was measured by ICP-MS (Agilent 7800, Agilent Technologies, Santa Clara, CA, USA), and the content of oxygen was measured by nitrogen & oxygen analyzer (ONH836, LECO Corporation, St. Joseph, MI, USA).

ZEISS EVO18 SEM-EDS (Carl Zeiss AG, Oberkochen, Germany) was used to analyze the morphology and composition of inclusions in steel, and the OPTON OTSInca system (OTSInca, OPTON Co., Ltd., Beijing, China) was used to analyze density, aspect ratio, and $D_{inclusion\ spacing}$ ($D_{inclusion\ spacing}$ represents the average distance between each inclusion and its nearest inclusion) of inclusions in the steel. In order to reveal the three-dimensional morphology of MnS inclusions, the samples were electrolytically etched in 1% tetramethylammonium chloride—10% acetylacetone—methanol [42], and the morphology was observed by SEM-EDS. FactSage 8.0 software (FactSage, Thermfact Ltd. (Montreal, QC, Canada) & GTT-Technologies (Aachen, Germany)) was used to analyze the formation of inclusions in liquid steel at 1873K and the precipitation of MnS during solidification. ZEISS ULTRA PLUS SEM-EDS (Carl Zeiss AG, Oberkochen, Germany) was used to analyze the surface and longitudinal section of the tensile fracture.

3. Results and Discussion

3.1. Steel Composition

The steel composition of Steel A–C was listed in Table 1, the yield of main alloy elements was stable. The sulfur content of Steel A–C was 0.01%, 0.06% and 0.132%, and represented low sulfur, medium sulfur and high sulfur respectively. It can be seen that the yield of sulfur of Steel A–C was 66.7%, 85.7% and 88%, respectively. As the increase of sulfur content, cerium content and magnesium content have a weak downward trend, but it can still be considered as the same level. In order to simulate production, ferroalloy used in the factory was used in this experiment, and calcium may come from calcium impurities in ferroalloy, which was illustrated in some literature [43,44].

Table 1. Steel composition (%).

Steel	C	Si	Mn	Cr	Ni	Cu	Mo	Als	Ca	Mg	Ce	T.O	S	Category
A	0.191	0.320	0.865	1.210	0.16	0.17	0.03	0.020	0.0019	0.0010	0.014	0.0026	0.01	Low sulfur
B	0.203	0.298	0.872	1.210	0.15	0.17	0.03	0.017	0.0022	0.0009	0.013	0.0029	0.06	Medium sulfur
C	0.214	0.313	0.869	1.225	0.15	0.17	0.03	0.016	0.0020	0.0008	0.013	0.0030	0.132	High sulfur

3.2. Evolution of Inclusions in Liquid Steel

Figure 4 lists the typical inclusion morphology in Sample S0 of Steel A–C. Figure 4a was MnS inclusion, which was mainly formed during the solidification process of steel. In the present study, although water quenching is conducted as soon as possible after Sample S0 was extracted, this did not completely avoid the formation of MnS. Figure 4b is a Ca-Mg-Al-Si-Mn-O-S composite inclusion, and this inclusion has a spherical shape, indicating that the inclusion may be liquid in steel. Some literature [45–47] has explained the formation mechanism of Ca-Mg-Al-O inclusion, which is mainly formed in the secondary refining process. In this experiment, SiO_2 formed in composite inclusion with the consumption of acid-soluble aluminum by slag-steel reaction during the period from 13 min to 58 min, and the Ca-Mg-Al-Si-O inclusion formed in steel in final. Figure 4c is a spinel inclusion, which was one of the typical inclusions in Al-killed steel.

Figure 5 shows typical inclusions detected in Sample S1–S2 of Steel A–C. In steel A, after the addition of Ce-Mg lump, it was found that Ce-Al-O inclusions and Mg-Al-O inclusions formed in Sample S1, as shown by A-S1-(a) and A-S1-(b). Cerium reacted with inclusions Ca-Mg-Al-Si-O and transformed into Ce-Ca-Mg-Al-Si-O inclusions, as demonstrated by A-S1-(c). A-S1-(d) showed a novel composite inclusion with a double layer structure: Ce-O-S inclusion was the core inclusion bearing with Ce-Ca-Mg-Al-Si-O inclusions. According to the detection results of Sample S2 of Steel A, the composition of inclusions changed with the continuation of the experiment. The composite inclusion bearing Ce-O-S inclusion (A-S1-(d)) disappeared, while Ce-Al-O inclusions, spinel, Ce-Ca-Mg-Al-Si-O inclusions can still be detected in Sample S2. Besides, (Ca, Mn) S also formed in steel, which is usually considered to be formed during solidification.

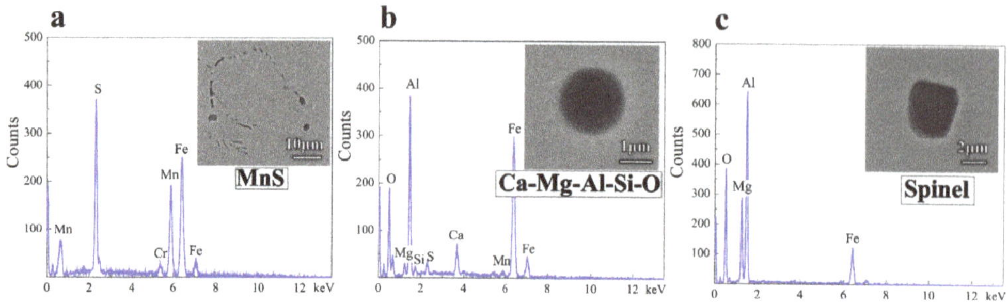

Figure 4. Typical morphology and composition of the inclusions observed in S0 sample. (**a**) MnS inclusion (**b**) Ca-Mg-Al-Si-O inclusion (**c**) Spinel inclusion.

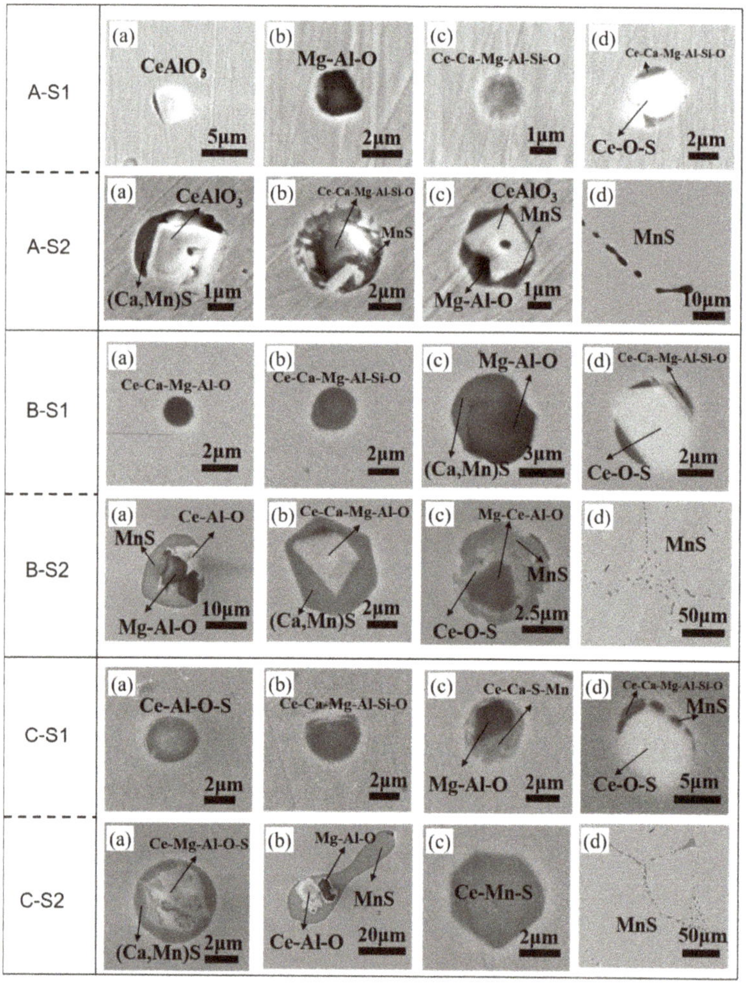

Figure 5. Typical morphology and classification of the inclusions observed in S1–S2 sample. A-S1-(**a**)–A-S1-(**d**) Inclusion in A-S1. A-S2-(**a**)–A-S2-(**d**) Inclusion in A-S2. B-S1-(**a**)–B-S1-(**d**) Inclusion in B-S1. B-S2-(**a**)–B-S2-(**d**) Inclusion in B-S2. C-S1-(**a**)–C-S1-(**d**) Inclusion in C-S1. C-S2-(**a**)–C-S2-(**d**) Inclusion in C-S2.

In Sample S1 of Steel B and Steel C, inclusion types were identical with Steel A. A similar inclusion type exhibited a composite inclusion with a double layer structure: Ce-O-S inclusion serving as a core wrapped by Ce-Ca-Mg-Al-Si-O inclusions can be also found during detection. Typical element mapping of this type of inclusion was shown in Figure 6. With the continuation of the experiment, inclusion with that structure disappeared. However, Ce-O-S inclusion still existed in the composite inclusion as a constituent phase in Sample S2, as shown in B-S2-(c) and C-S2-(a), and this was the main difference from Steel A. In addition, Ce-Mn-S inclusions with a low aspect ratio can be detected only in Steel C, as shown in C-S2-(c).

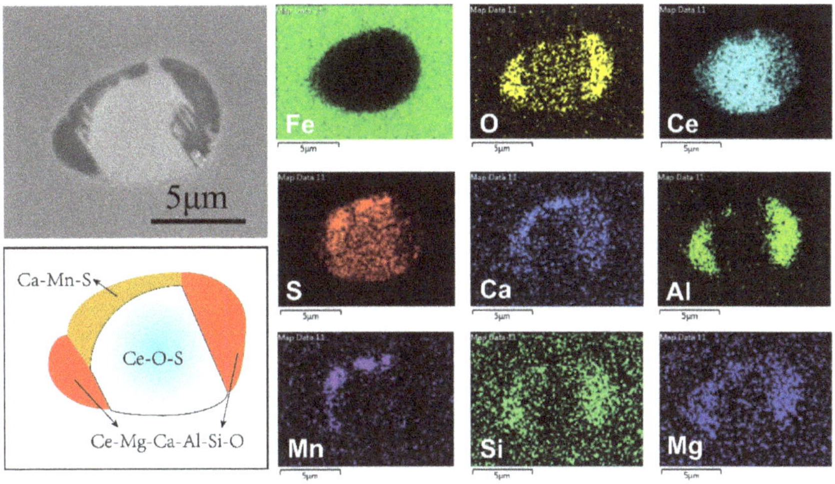

Figure 6. Typical elemental mapping of transition state inclusion in S1 sample for experiments A–C.

Figure 7 presents the proportion of MnS inclusions in Steel A–C. It can be seen that with the increase of sulfur content, the number of MnS inclusions in as-cast samples increased, and the proportion of MnS inclusions in Steel B and C both exceeded 90%, which was the primary inclusions. Figure 8 shows density, aspect ratio and $D_{inclusion\ spacing}$ of inclusion. With the increase of sulfur content, inclusions density increased, aspect ratio increased, but the inclusion spacing decreased, and their relationship was approximately linear. It indicated that the sulfur content has a significant impact on the MnS precipitation and morphology.

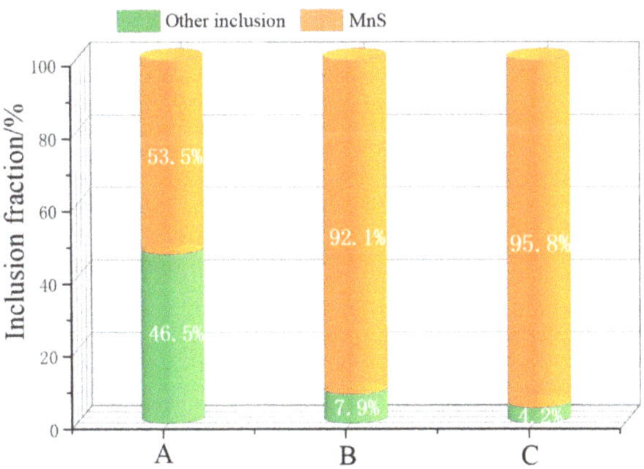

Figure 7. Percentage distribution of inclusion classification in S2 sample for Steel A–C.

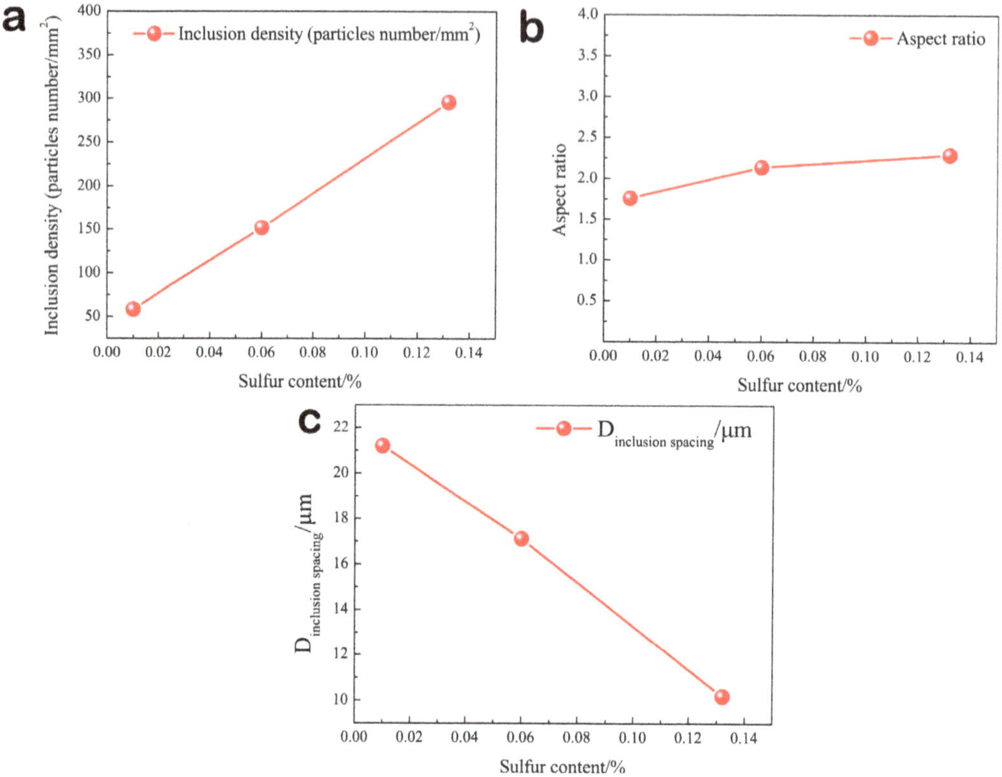

Figure 8. Inclusion density, aspect ratio and $D_{inclusion\ spacing}$ in Sample S2 of Steel A–C. (**a**) Inclusion density. (**b**) Aspect ratio. (**c**) $D_{inclusion\ spacing}$.

A summary of the phenomenon that the sulfur content has an impact on inclusion evolution in liquid steel at 1873K and MnS inclusion precipitation during solidification can be obtained as follows, a type of double structure inclusion that Ce-O-S inclusion serving as a core wrapped by Ce-Ca-Mg-Al-Si-O inclusions can form in liquid steel after Ce-Mg lump addition and disappeared with experiment time prolonging. In Sample S2, Ce-O-S inclusion can not be detected in Steel A but can be found in Steel B and Steel C. The increase of sulfur content promotes MnS precipitation and also has an impact on the $D_{inclusion\ spacing}$ and morphology of MnS.

3.3. Discussion on Evolution Mechanism of Inclusions in Liquid Steel

To illustrate the inclusion mechanism in liquid steel after Ce-Mg lump addition, a thermodynamic calculation was conducted by FactSage software. Database was selected as FactPS and FSstel, and the system was 0.20C-0.30Si-0.87Mn-0.02Al-0.001Mg-0.0020Ca-0.003O-1.20 Cr-S-Ce-Fe (balance). It can be seen from Figure 9, the results showed that with the increase of cerium content, the evolution sequence of Ce-containing inclusions in 0.01% sulfur content steel was $CeAlO_3 \rightarrow Ce_2O_3 \rightarrow Ce_2O_2S$, and the evolution sequence of Ce-containing inclusions in 0.06% and 0.132% sulfur content steel was $CeAlO_3 \rightarrow Ce_2O_2S \rightarrow Ce_2S_3$. When cerium content was fixed as 0.013%, the variation of cerium-containing inclusions was $CeAlO_3 \rightarrow Ce_2O_2S$ with the increase of sulfur content. From here, it can be concluded that the increase of cerium content and sulfur content can both result in the transition between $CeAlO_3$ and Ce_2O_2S.

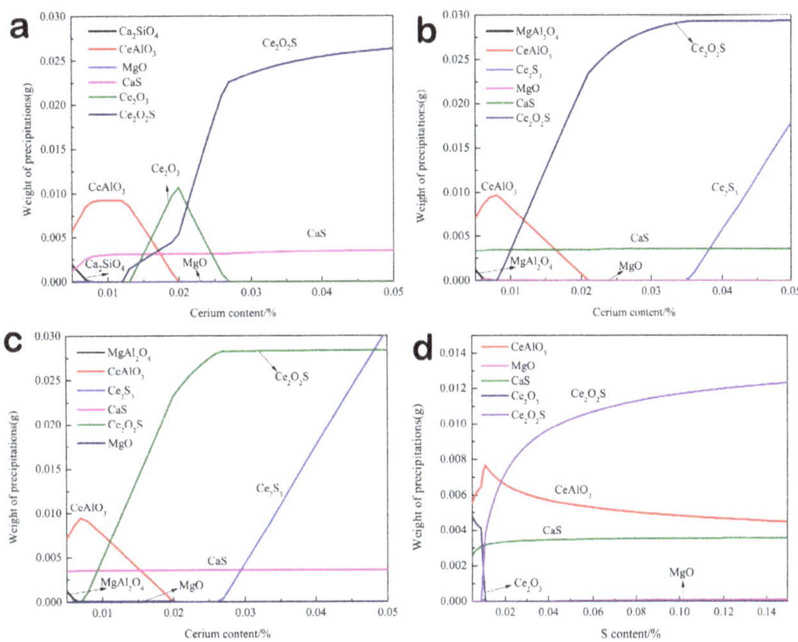

Figure 9. Thermodynamic calculation of the effect of cerium content and sulfur content on inclusion formation by FactSage. (**a**) Effect of Ce content on inclusion formation in steel containing 0.01% sulfur content (**b**) Effect of Ce content on inclusion formation in steel containing 0.06% sulfur content (**c**) Effect of Ce content on inclusion formation in steel containing 0.132% sulfur content (**d**) Effect of S content on inclusion formation in steel containing 0.013% Ce content.

Based on experimental results and thermodynamic calculation results, the inclusion evolution mechanism for this research was concluded as shown in Figure 10. In a short time after adding the Ce-Mg lump, due to the limitation of mass transfer, there was a concentrated area of cerium in liquid steel. Considering thermodynamic calculations, the high content of cerium in the concentrated area resulted in the formation of Ce_2O_2S. At the same time, the addition of magnesium can bring a severe stirring of the molten pool because the solubility of magnesium in liquid steel is very low, resulting in the probability of physical collision of inclusion increasing greatly. Under this system conditions, a comprehensible situation is that Ce_2O_2S, as a high melting point inclusion, presents a solid-state in liquid steel, and collides with Ca-Mg-Al-Si-O inclusion in a liquid state. As a result, the liquid phase adhered to the outside of the solid phase and formed a composite inclusion with a double-layer structure, as shown in A-S1-(d), B-S1-(d), C-S1-(d) in Figure 5. Considering that formation mechanism and experimental phenomenon, this inclusion was named transition state inclusion. With the homogenization of steel and conduct of chemical reactions, in Steel A, due to the $CeAlO_3$ has higher formation ability than Ce_2O_2S, Ce_2O_2S phase of transition state inclusion transition into $CeAlO_3$ as expressed in reaction (1).

$$Ce_2O_2S + 2[Al] + 4[O] = 2CeAlO_3 + [S] \quad (1)$$

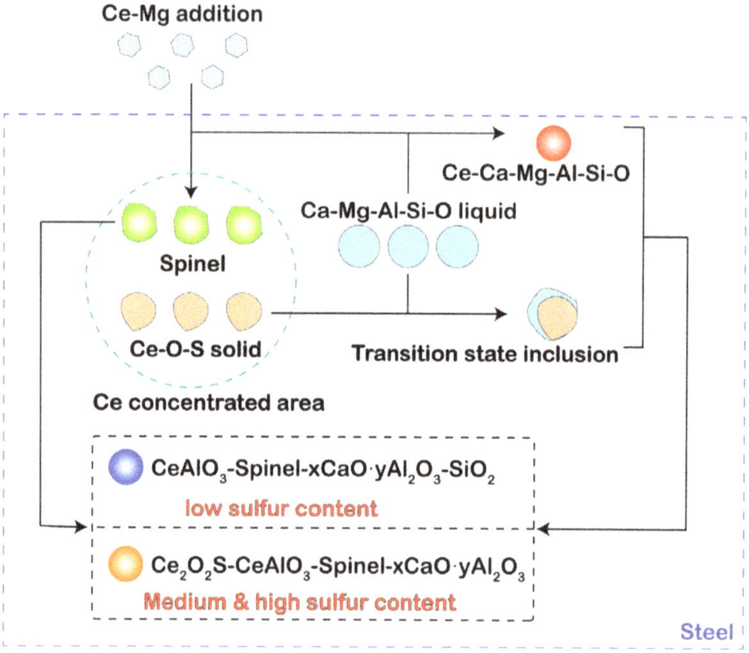

Figure 10. Schematic diagram of inclusion evolution after Ce-Mg addition.

In Steel B and Steel C, the formation ability of Ce_2O_2S is stronger than $CeAlO_3$. Therefore, Ce-O-S of transition state inclusion was not replaced, and as the substitution reaction conducting at the outer layer, SiO_2 and CaO in the outer layer were consumed according to reaction (2)–(5), which can be seen as B-S2-(c) and C-S2-(a).

$$4[Ce] + 2(Al_2O_3)_{inclusion} + 3(SiO_2)_{inclusion} = 4CeAlO_3 + 3[Si] \quad (2)$$

$$2[Ce] + (Al_2O_3)_{inclusion} + 3(CaO)_{inclusion} = 2CeAlO_3 + 3[Ca] \quad (3)$$

$$2[Ce] + (SiO_2)inclusion + [S] = Ce_2O_2S + [Si] \quad (4)$$

$$2[Ce] + 2(CaO)_{inclusion} + [S] = Ce_2O_2S + 2[Ca] \quad (5)$$

Also, based on the experimental phenomena, there were two reactions conducted in liquid steel as follows, (1) Cerium reacted directly with aluminum and oxygen to form $CeAlO_3$ in liquid steel. (2) Cerium reacted with the Ca-Mg-Al-O inclusion forming before adding Ce-Mg lump with the consumption of the unstable oxide components (such as SiO_2) and forming composite inclusion Ce-Ca-Mg-Al-Si-O.

3.4. Precipitation of MnS Inclusion during Solidification

The three-dimensional morphology of MnS observed by SEM was presented in Figure 11a–l. It is obviously that the size of MnS inclusion has become much more extensive and the morphology of MnS has become from "stick" to "dendritic + fishbone" to "fishbone" with the increase of sulfur content. Especially, MnS inclusions showed a multi-directional growth morphology in Steel C, which is consistent with the analysis in two-dimensional morphology.

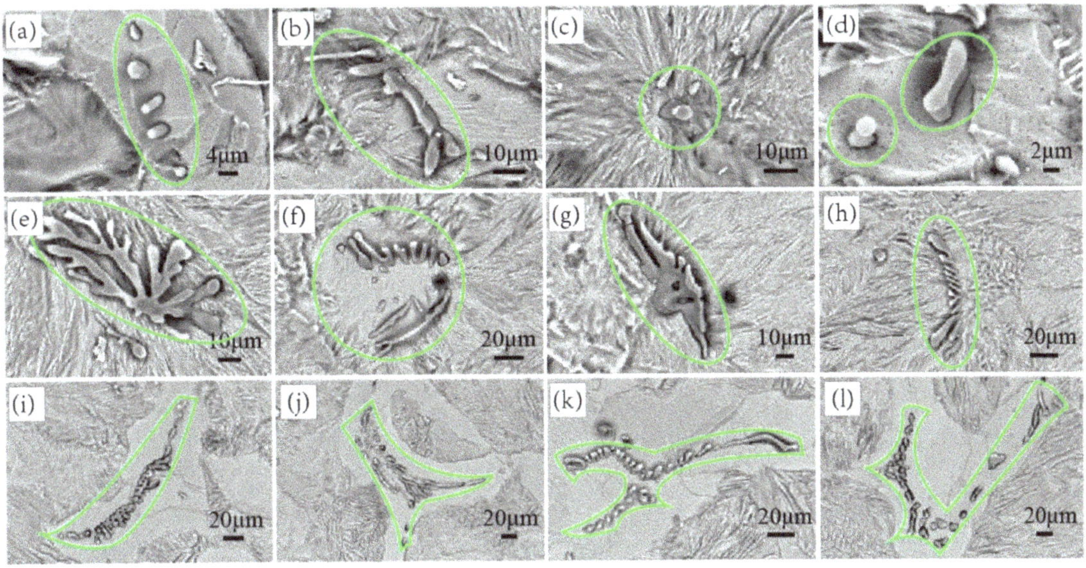

Figure 11. 3D morphology of MnS inclusions in as-cast sample. (**a–d**) Sample S2 of Steel A; (**e–h**) Sample S2 of Steel B; (**i–l**) Sample S2 of Steel C.

For understanding the precipitation behavior of MnS, in the present study, FactSage software was used to analyze the precipitation behavior of MnS during solidification under different sulfur contents. The calculation was based on the Scheil-Gulliver model [48] which assumes that no diffusion takes place in the solid and that solute redistribution in the liquid is fast. The database were FactPS and FSstel, and the calculation system is 0.20C-0.30Si-0.87Mn-0.02Al-0.001Mg-0.0020Ca-0.003O-1.20Cr-S-0.013Ce-Fe(balance). The calculation results were shown in Figure 12. With the increase of sulfur content, the temperature point at which MnS precipitated in large quantities advances during solidification. The increase of sulfur content resulted in the increase of the mass fraction of MnS obviously, and when the sulfur content reached 0.132%, the mass fraction of MnS in the system reached 0.21%.

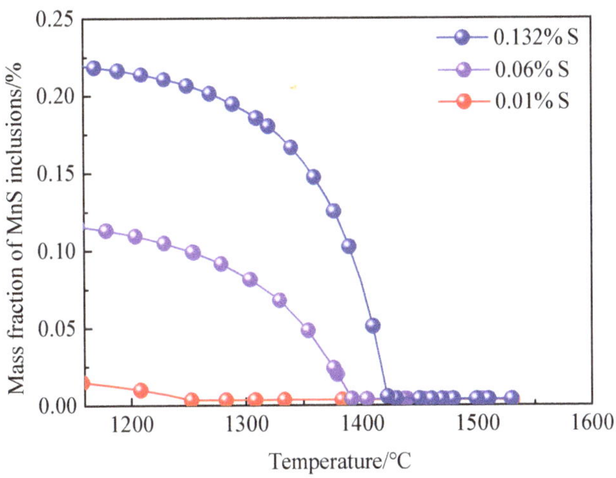

Figure 12. Thermodynamic calculation of MnS inclusion formation during solidification process by FactSage.

3.5. Effects of MnS Inclusions on Mechanical Properties

Figure 13 presents the tensile strength and elongation of Steel A–C. It can be seen that with the increase of sulfur content, tensile strength and elongation showed a downward trend. Tensile strength decreased from 1443.8 MP to 1343.57 MPa, and the elongation decreased from 9.7% to 6.8%. The negative effect of the increase of sulfur content on the strength and ductility of steel was noticeable.

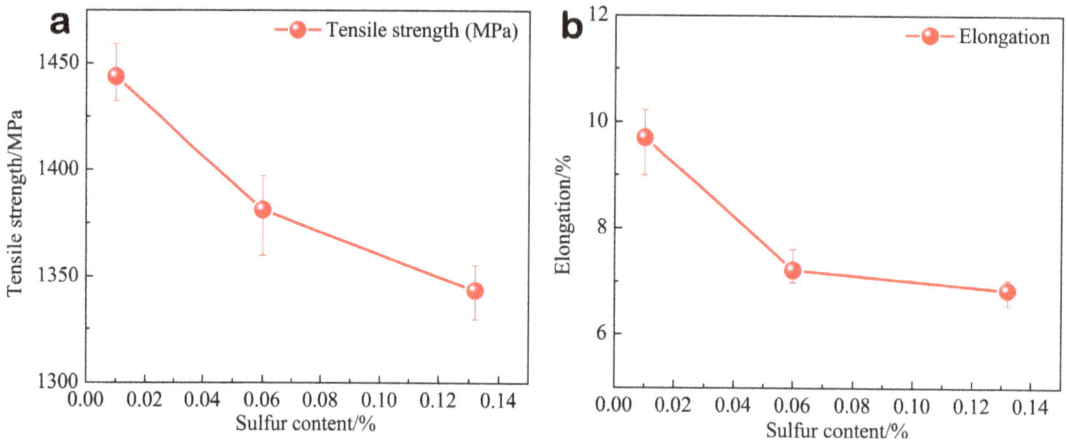

Figure 13. Tensile strength and elongation result. (**a**) Tensile strength; (**b**) Elongation.

To explore the impact mechanism of sulfur content on tensile fracture, the fracture morphology of Steel A–C was observed and analyzed. Representative fractography of tensile specimens was presented in Figure 14. According to Figure 14a–i, the fracture morphology consisted of major cracks, voids, and dimples. With the increase of sulfur content, the void amount increased obviously, and the morphology that multiple voids coalescenced into large-sized voids can be observed clearly in Steel C as shown in Figure 14i. MnS inclusion can be found in some voids, which indicates that MnS inclusion provided a nucleation point for void initiation. Figure 14j–l present the observation and analysis of the longitudinal section of the tensile fracture. It can be seen that MnS inclusion located at the root of a void. It can be inferred that stress concentration occurs at the tip of MnS, which resulted in void initiation. This is consistent with the research of Yamamoto et al. [49].

The fracture of steel was usually illustrated as void nucleation, growth, coalescence. As the nucleation point of the void, the distance between inclusions may greatly determine the difficulty of void coalescence. In the present study, the increase of sulfur content caused the decline of inclusion spacing and resulted in a shorter physical path of crack propagation in steel, which makes void coalescence occurrence easier. A schematic diagram of this mechanism was presented in Figure 14m.

For the purpose of machinability enhancement, the sulfur content of gear steel SCr420H is considered to increase in the future. The volume of MnS will increase a lot in steel because of the increase of sulfur content. Excessive MnS inclusion can increase the nucleation points of cracks inevitably and bring a harmful effect on the mechanical properties of steel. Therefore, it is essential to adopt some metallurgical means to make MnS distribute uniformly, and the damage of inclusions to mechanical properties can be suppressed.

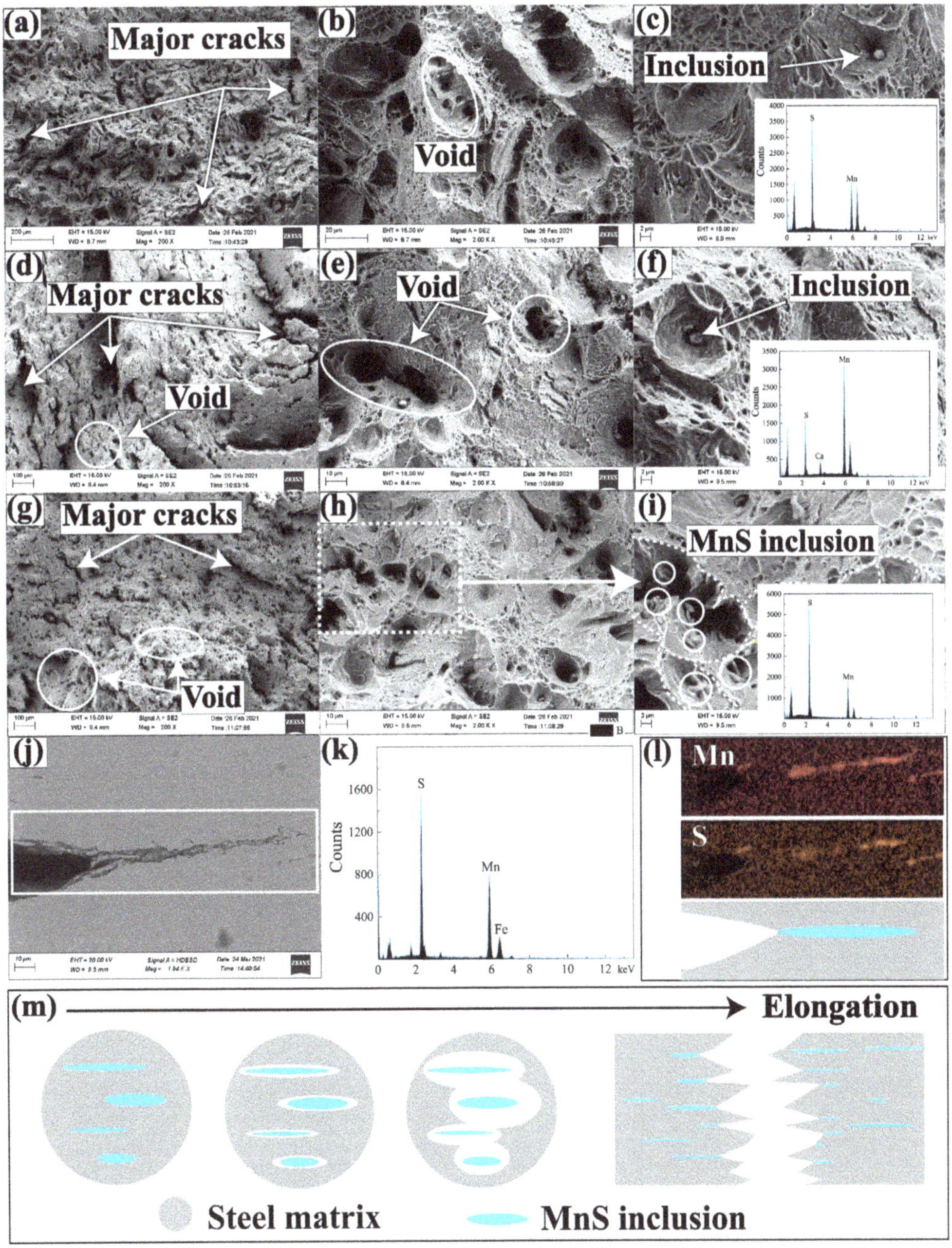

Figure 14. Morphology observation and inclusion analysis on tensile fracture. (**a–c**) tensile fracture of Steel A; (**d–f**) tensile fracture of Steel B; (**g–i**) tensile fracture of Steel C; (**j–l**) Longitudinal section of tensile fracture of Steel C; (**m**) Schematic diagram of MnS inclusion initiation fracture.

4. Conclusions

The current work investigated the effect of sulfur content on inclusion and mechanical properties by laboratory experiment and thermodynamic calculation. The main conclusions were summarized as follows.

(1) After the Ce-Mg addition, the cerium concentration area formed in liquid steel. This may result in the formation of transition state inclusions. In current experiment, solid Ce_2O_2S formed in the cerium concentration area and collides with Ca-Mg-Al-Si-O liquid inclusion, and the liquid phase adhered to the outside of the solid phase, and formed a composite inclusion with a double-layer structure. As a type of transition state inclusion, it only exists in a short time after Ce-Mg addition. As the homogenization of liquid steel, cerium concentration area disappeared, and the difference of sulfur content in steel can result in this type of transition state inclusions transformed into different inclusions.

(2) The thermodynamic calculation results indicated that with the increase of cerium content, the evolution sequence of Ce-containing inclusions in 0.01% sulfur content steel is $CeAlO_3 \rightarrow Ce_2O_3 \rightarrow Ce_2O_2S$, and the evolution sequence of Ce-containing inclusions in 0.06% and 0.132% sulfur content steel is $CeAlO_3 \rightarrow Ce_2O_2S \rightarrow Ce_2S_3$.

(3) With the sulfur content increasing from 0.01% to 0.132%, MnS becomes the prominent inclusion in steel, and inclusion spacing decreases from 21.20 μm to 10.19 μm, the tensile strength decreases from 1443.8 MPa to 1343.57 MPa, and the elongation decreases from 9.7% to 6.8%.

Author Contributions: Conceptualization, M.S. and Z.J.; Methodology, M.S. and Y.L.; Formal analysis, M.S. and S.M.; Writing, M.S., Y.J. and J.W.; Resources, C.C. and H.L. All authors have read and agreed to the published version of the manuscript.

Funding: This work was supported by the National Natural Science Foundation of China N. 52074075 and the National Key Research and Development Program of China N. 2016YFB0300105.

Institutional Review Board Statement: Not applicable.

Informed Consent Statement: Not applicable.

Data Availability Statement: Not applicable.

Conflicts of Interest: The authors declare no conflict of interest.

References

1. Xie, J.B.; Zhang, D.; Yang, Q.K.; An, J.M.; Huang, Z.Z.; Fu, J.X. Exploration of morphology evolution of the inclusions in Mg-treated 16MnCrS5 steel. *Ironmak. Steelmak.* **2019**, *46*, 564–573. [CrossRef]
2. Sun, H.; Wu, L.P.; Xie, J.B.; Ai, K.N.; Zeng, Z.Q.; Shen, P.; Fu, J.X. Inclusions modification and improvement of machinability in a non-quenched and tempered steel with Mg treatment. *Metall. Res. Technol.* **2020**, *117*, 208. [CrossRef]
3. Ånmark, N.; Karasev, A.; Jönsson, P.G. The effect of different non-metallic inclusions on the machinability of steels. *Materials* **2015**, *8*, 751–783. [CrossRef]
4. Sui, H.; Wang, L.J.; Wang, Q.; Wang, H.M.; Che, D.H.; Li, J.M. The formation and growth of sulfides in free-cutting stainless steel. *Steel Res. Int.* **2018**, *89*, 1800179. [CrossRef]
5. Ånmark, N.; Karasev, A.; Jönsson, P.G. The influence of microstructure and non-metallic inclusions on the machinability of clean steels. *Steel Res. Int.* **2017**, *88*, 1600111. [CrossRef]
6. An, X.X.; Tian, Y.; Wang, H.J.; Shen, Y.F.; Wang, Z.D. Suppression of austenite grain coarsening by using Nb-Ti microalloying in high temperature carburizing of a gear steel. *Adv. Eng. Mater.* **2019**, *21*, 1900132. [CrossRef]
7. An, X.X.; Tian, Y.; Wang, H.J.; Wang, Z.D. Effect of preheat treatment on microstructure and properties of a gear steel for high-temperature carburizing. *Steel Res. Int.* **2020**, *91*, 2000180. [CrossRef]
8. Wen, B.; Song, B.; Pan, N.; Hu, Q.Y.; Mao, J.H. Effect of austenitizing temperature on microstructure in 16Mn steel treated by cerium. *Int. J. Miner. Metall. Mater.* **2011**, *18*, 652–658. [CrossRef]
9. Ji, Y.P.; Zhang, M.X.; Ren, H.P. Roles of lanthanum and cerium in grain refinement of steels during solidification. *Metals* **2018**, *8*, 884. [CrossRef]
10. Ji, Y.P.; Li, Y.M.; Zhang, M.X.; Qu, W.; Zhao, T.X.; Ren, H.P. Grain refinement mechanism of the δ-ferrite in steels through cerium addition. *Metall. Mater. Trans. A* **2020**, *51*, 1707–1718. [CrossRef]
11. Jiang, X.; Song, S.H. Enhanced hot ductility of a Cr-Mo low alloy steel by rare earth cerium. *Mater. Sci. Eng. A* **2014**, *613*, 171–177. [CrossRef]
12. Gao, J.Z.; Fu, P.X.; Liu, H.W.; Li, D.Z. Effects of rare earth on the microstructure and impact toughness of H13 steel. *Metals* **2015**, *5*, 383–394. [CrossRef]
13. Jiang, Z.H.; Wang, C.; Gong, W.; Wang, H.D. Evolution of inclusions and change of as-cast microstructure with Mg addition in high carbon and high chromium die steel. *Ironmak. Steelmak.* **2015**, *42*, 669–674. [CrossRef]

14. Cui, X.K.; Song, B.; Yang, Z.B.; Liu, Z.; Li, L.F.; Wang, L. Effect of Mg on the evolution of inclusions and formation of acicular ferrite in La-Ti-treated steels. *Steel Res. Int.* **2020**, *91*, 1900563. [CrossRef]
15. Jiang, Z.H.; Xu, G.; Li, Y.; Li, H.B.; Lv, J.B.; Wang, Q. Effect of ultra-high magnesium on SKS51 liquid steel cleanliness and microstructure. *ISIJ Int.* **2019**, *59*, 1234–1241. [CrossRef]
16. Bao, D.H.; Cheng, G.G.; Huang, Y.; Qiao, T.; Dai, W.X. Refinement of Solidification Structure of H13 Steel by Rare Earth Sulfide. *Steel Res. Int.* **2021**, *93*, 2100304. [CrossRef]
17. Shin, J.H.; Park, J.H. Formation mechanism of oxide-sulfide complex inclusions in high-sulfur containing steel melts. *Metall. Mater. Trans. A* **2017**, *49*, 311–324. [CrossRef]
18. Li, X.; Long, X.; Wang, L.; Tong, S.; Wang, X.; Zhang, Y.; Li, Y. Inclusion characteristics in 95CrMo steels with different calcium and sulfur contents. *Materials* **2020**, *13*, 619. [CrossRef]
19. Zeng, J.; Zhu, C.Y.; Wang, W.L.; Li, X. In situ observation of the MnS precipitation behavior in high-sulfur microalloyed steel under different cooling rates. *Metall. Mater. Trans. B* **2020**, *51*, 2522–2531. [CrossRef]
20. Maciejewski, J. The effects of sulfide inclusions on mechanical properties and failures of steel components. *J. Failure Anal. Prev.* **2015**, *15*, 169–178. [CrossRef]
21. Huang, Y.; Cheng, G.G.; Li, S.J.; Dai, W.X. Effect of cerium on the behavior of inclusions in H13 steel. *Steel Res. Int.* **2018**, *89*, 1800371. [CrossRef]
22. Ren, Q.; Zhang, L.F. Effect of cerium content on inclusions in an ultra-low-carbon aluminum-killed steel. *Metall. Mater. Trans. B* **2020**, *51*, 589–600. [CrossRef]
23. Hamidzadeh, M.A.; Meratian, M.; Saatchi, A. Effect of cerium and lanthanum on the microstructure and mechanical properties of AISI D2 tool steel. *Mater. Sci. Eng. A* **2013**, *571*, 193–198. [CrossRef]
24. Lin, C.K.; Pan, Y.C.; Su, Y.H.F.; Lin, G.R.; Hwang, W.S.; Kuo, J.C. Effects of Mg-Al-O-Mn-S inclusion on the nucleation of acicular ferrite in magnesium-containing low-carbon steel. *Mater. Charact.* **2018**, *141*, 318–327. [CrossRef]
25. Du, G.; Li, J.; Wang, Z.B.; Shi, C.B. Effect of magnesium addition on behavior of collision and agglomeration between solid inclusion particles on H13 steel meltes. *Steel Res. Int.* **2017**, *88*, 1600185. [CrossRef]
26. Wang, H.; Li, J.; Shi, C.B.; Li, J. Eolution of Al_2O_3 inclusions by magnesium treatment in H13 hot work die steel. *Ironmak. Steelmak.* **2016**, *44*, 128–133. [CrossRef]
27. Huang, Y.; Cheng, G.G.; Xie, Y. Modification Mechanism of Cerium on the Inclusions in Drill Steel. *Acta Metall. Sin.* **2018**, *54*, 1253–1261.
28. Roos, E.; Karasev, A.; Jönsson, P.G. Effect of Si and Ce contents the nozzle clogging in a REM alloyed stainless steel. *Steel Res. Int.* **2015**, *86*, 1279–1288. [CrossRef]
29. Shen, P.; Fu, J. Morphology study on inclusion modification using Mg-Ca treatment in resulfurized special steel. *Materials* **2019**, *12*, 197. [CrossRef] [PubMed]
30. Li, X. Metallurgical Quality and Composition Optimization of H13 Hot Work Die Steel. Ph.D. Thesis, Northeastern University, Shenyang, China, 2020.
31. Luyckx, L.; Bell, J.R.; Mclean, A.; Korchynsky, M. Sulfide Shape Control in High Strength Low Alloy Steels. *Metall. Trans.* **1970**, *1*, 3341–3350.
32. Dahl, W.; Gammal, T.E.; Lorenz, L.L. Einfluß sehr niedriger Schwefelgehalte auf die mechanischen Eigenschaften des Stahles St 52-3. *Arch. Eisenhüttenwesen* **1973**, *44*, 843–846. [CrossRef]
33. Banks, T.M.; Gladman, T. Sulphide shape control. *Met. Technol.* **1979**, *6*, 81–94. [CrossRef]
34. Handerhan, K.J.; Garrison, W.M.; Moody, N.R. A Comparison of the Fracture Behavior of Two Heats of the Secondary Hardening Steel AF1410. *Metall. Trans. A* **1989**, *20*, 105–123. [CrossRef]
35. Li, X.; Jiang, Z.H.; Geng, X.; Chen, M.J.; Cui, S. Effect of rare earth-magnesium alloy on inclusion evolution in industrial production of die steel. *Steel Res. Int.* **2019**, *90*, 1900103. [CrossRef]
36. Li, X.; Jiang, Z.H.; Geng, X.; Chen, M.J.; Peng, L.Z. Evolution mechanism of inclusions in H13 steel with rare earth magnesium alloy addition. *ISIJ Int.* **2019**, *59*, 1552–1561. [CrossRef]
37. Chang, L.Z.; Gao, G.; Zheng, F.Z.; Shi, X.F. Effect of rare earth and magnesium complex treatment on inclusions in GCr15 bearing steel. *J. Chem. Eng.* **2019**, *41*, 763–771.
38. Maloney, J.L.; Garrison, W.M. The effect of sulfide type on the fracture behavior of HY180 steel. *Acta Mater.* **2005**, *53*, 533–551. [CrossRef]
39. Ma, Y.; Pan, T.; Jiang, B.; Cui, Y.H.; Su, H.; Peng, Y. Study of the effect of sulfur contents on frature toughness on railway wheel steels for high speed train. *Acta Metall. Sin.* **2011**, *47*, 978–983.
40. Yang, C.Y.; Luan, Y.K.; Li, D.Z.; Li, Y.Y. Very hgih cycle fatigue properties of bearing steel with different aluminum and sulfur content. *Int. J. Fatigue* **2018**, *116*, 396–408. [CrossRef]
41. Kang, M.H.; Lee, J.S.; Koo, Y.M.; Kim, S.J.; Heo, N.H. Correlation between MnS precipitation, sulfur segregation kinetics, and hot ductility in C-Mn steel. *Metall. Mater. Trans. A* **2014**, *45*, 5295–5299. [CrossRef]
42. Oikawa, K.; Ohtani, H.; Ishida, K.; Nishizawa, T. The control of the morphology of MnS inclusions in steel during solidification. *ISIJ Int.* **1995**, *4*, 402–408. [CrossRef]
43. Wijk, O.; Brabie, V. The purity of ferrosilicon and its influence on inclusion cleanliness of steel. *ISIJ Int.* **1996**, *36*, 132–135. [CrossRef]

44. Mizuno, K.; Todoroki, H.; Noda, M.; Tohge, T. Effects of Al and Ca in ferrosilicon alloys for deoxidation on inclusion composition in type 304 stainless steel. *Iron Steelmak.* **2001**, *28*, 93–101.
45. Deng, Z.Y.; Zhu, M.Y. Evolution mechanism of non-metallic inclusions in Al-killed alloyed steel during secondary refining process. *ISIJ Int.* **2013**, *53*, 450–458. [CrossRef]
46. Liu, C.Y.; Huang, F.X.; Wang, X.H. The effect of refining slag and refractory on inclusion transformation in extra low oxygen steels. *Metall. Mater. Trans. B* **2016**, *47*, 999–1009. [CrossRef]
47. Jiang, M.; Wang, X.H.; Chen, B.; Wang, W.J. Laboratory study on evolution mechanisms of non-metallic inclusions in high strength alloyed steel refined by high basicity slag. *ISIJ Int.* **2010**, *50*, 95–104. [CrossRef]
48. Gulliver, G.H. The quantitative effect of rapid cooling upon the constitution of binary alloys. *J. Inst. Met.* **1913**, *9*, 120–157.
49. Yamamoto, K.I.; Yamamura, H.; Suwa, Y. Behavior of non-metallic inclusions in steel during hot deformation and the effects of deformed inclusions on local ductility. *ISIJ Int.* **2011**, *12*, 1987–1994. [CrossRef]

metals MDPI

Article

Effects of Rare Earth La–Ce Alloying Treatment on Modification of Inclusions and Magnetic Properties of W350 Non-Oriented Silicon Steel

Haijun Wang [1], Yuhao Niu [1], Haitao Ling [1,*], Jialong Qiao [1,2], Yanling Zhang [3], Wei Zhong [4] and Shengtao Qiu [2]

1 Anhui Province Key Laboratory of Metallurgical Engineering & Resources Recycling, Anhui University of Technology, Maanshan 243002, China
2 National Engineering Research Center of Continuous Casting Technology, China Iron & Steel Research Institute Group, Beijing 100081, China
3 State Key Laboratory of Advanced Metallurgy, University of Science and Technology Beijing, Beijing 100083, China
4 Xinyu Iron and Steel Co., Ltd., Xinyu 338001, China
* Correspondence: linghaitao@ahut.edu.cn

Abstract: In order to study the effects of rare earth La–Ce alloying treatment on the characteristics of inclusions in non-oriented silicon steels, industrial experiments were conducted studying the composition, morphology, size and quantity of inclusions in W350 non-oriented silicon steel during the RH (Ruhrstahl-Hereaeus) refining process and tundish process, after rare earth treatment. The products were analyzed by means of ICP-MS (inductively coupled plasma mass spectrometry), SEM/EDS (scanning electron microscope-energy dispersive spectrometry), and ASPEX (automated SEM/EDS inclusion analysis). The research results showed that the types of inclusions in experimental steel changed significantly after rare earth treatment. The types of inclusions after RE (rare earth) treatment are typically rare earth composite inclusions that are mainly composed of (La, Ce)Al_2O_3, and conventional inclusions. The addition of rare earth promotes the agglomeration of inclusions; the morphologies of the inclusions are mostly blocky, and some are distributed in long strips. After rare earth treatment during the RH refining process, the number of inclusions with sizes of 1.0~3.5 μm in the experimental steel is increased, and the average size of the inclusions is 2.66 μm. In addition, the number of inclusions larger than 4 μm in the specimens increases due to the collision and growth of inclusions caused by the RH circulation. After rare earth treatment during the tundish process, the number of micro inclusions with sizes of 1.0~2.5 μm in the specimen steels decreases, while the number of inclusions larger than 5 μm increases. The size distribution of micro inclusions in hot-rolled sheets after rare earth treatment was studied using TEM (transmission electron microscopy). In the specimens without rare earth, the content of micro inclusions (≤1 μm) is 51,458.2/mm^2 and the average size is 0.388 μm. In the specimens with rare earth added, the content of micro inclusions (≤1 μm) is 24,230.2/mm^2 and the average size is 0.427 μm. Compared to sheet produced by the original process, the iron loss of the 0.35 mm finished experimental sheet is reduced by 0.068 W/kg, and the magnetic induction is increased by 0.007 T. The iron loss of the 0.50 mm finished experimental sheet is reduced by 0.008 W/kg, and the magnetic induction is increased by 0.004 T. After rare earth treatment, the average size of micro inclusions increases and the magnetic properties are obviously improved.

Keywords: rare earth La–Ce; inclusions; modification; non-oriented silicon steel; magnetic property

Citation: Wang, H.; Niu, Y.; Ling, H.; Qiao, J.; Zhang, Y.; Zhong, W.; Qiu, S. Effects of Rare Earth La–Ce Alloying Treatment on Modification of Inclusions and Magnetic Properties of W350 Non-Oriented Silicon Steel. *Metals* **2023**, *13*, 626. https://doi.org/10.3390/met13030626

Academic Editor: Jiro Kitagawa

Received: 30 January 2023
Revised: 17 March 2023
Accepted: 17 March 2023
Published: 21 March 2023

1. Introduction

High-grade non-oriented silicon steel, as the functional material for high-end power equipment [1,2], is widely used in the cores of large generators and high-efficiency energy-saving motors, in high-efficiency energy-saving appliances, and in electric vehicle manufacturing, due to its magnetic characteristics of low iron loss and high magnetic induction [3–5].

With the implementation and promotion of frequency conversion technologies in the home appliance industry, the rise of the new energy automobile industry and the development of energy-saving and high-efficiency motors, there is an increased demand for the low iron loss and high magnetic induction properties of high-grade non-oriented silicon steel [6]. Inclusions in steel are one of the many factors affecting its magnetic properties, in particular, the size of the inclusions. Since coercivity is inversely proportional to the size of inclusions, when the size of the inclusions is close to the thickness of the domain wall, the ability of inclusions to pin magnetic domains is usually at its strongest, and the deterioration of magnetic properties is typically at its most serious [7,8]. Relevant production enterprises and research institutes have carried out research on the modification and purification of inclusions in silicon steels, with the mechanisms and production applications of rare earth modified inclusions attracting a large number of researchers. Rare earth elements have unique metallurgical and physical properties, and can change the inherent types and morphologies of inclusions, reduce the harm of inclusions in the steel, and improve the related properties of steel products [9–14]. Takashima et al. [15,16] proposed the addition of composite rare earth and Al to non-oriented silicon steel. Their results showed that after adding rare earth alloys and Al, the inclusion sizes increased, the grain growth rate was significantly improved, the residual stress after annealing was reduced, and the product performance was significantly improved. Li [17], Kong [18], Qiao [19], and Wan [20–22] have studied the applications of rare earth elements in silicon steels. With the addition of rare earth elements, the magnetic properties of silicon steel products are obviously improved. The modification mechanism of rare earth elements on inclusions is relatively clear. However, there is a lack of relevant research on how the type and size distributions of inclusions formed in each process change after the addition of rare earth, especially the type transformations and size changes of micro inclusions in liquid steel.

In this work, industrial trials were carried out using La and Ce rare earth alloys during the production of silicon steels. The effects of rare earth La and Ce alloys on the evolution characteristics and metamorphism of inclusions, and the magnetic properties of products from different processes were investigated, which was conducive to realizing the stable application of rare earth elements in industrial production.

2. Materials and Methods

Slag was completely removed before the hot metal entered the desulfurization station, and slag was removed twice after the deep desulfurization treatment, to achieve $[S] \leq 0.0010\%$ and minimize the amounts of high-sulfur slag entering the converter. The scrap steel was self-produced high-quality scrap steel, of which about 50% was silicon steel scrap. During the tapping of the converter, the slide plate and the slag stopper were used to control the slag, and the thickness of the top slag in the ladle was required to be less than 60 mm. Lime was added to the top slag during the tapping process, and the argon flow rate was controlled to prevent the top slag of the ladle from agglomerating. During the RH refining process, it was forbidden to add aluminum to the liquid steel to melt it after the liquid steel had arrived at the RH refining station. After decarburization, the oxygen content of the liquid steel was less than 300 ppm, before alloying. After adding the desulfurizer and bauxite, rare earth alloys were added to liquid steel. The vacuum was broken after 8 min of net circulation and the sedation time was not less than 15 min.

The chemical composition of La and Ce alloys used in the experiment is listed in Table 1. Spectrography and inductively coupled plasma mass spectrometry (ICP-MS, Thermo Fisher Scientific (China) Co., Ltd., China) were used to determine the steel composition. The chemical composition of the experimental steel is shown in Table 2.

Table 1. Chemical composition of rare earth La–Ce alloys (wt%).

Element	C	Mg	Si	Ca	La	Ce
Content	0.002	0.001	0.2	0.001	37.32	61.67

Table 2. Chemical composition of experiment steel (wt%).

Element	C	Mn	S	P	Si	Als	N	Ti	Cu	La	Ce
RE-free (Rare earth-free)	0.0024	0.32	0.0022	0.016	2.69	0.53	0.001	0.0018	0.038	/	/
RE-added (Rare earth-added)	0.0026	0.36	0.0028	0.015	2.62	0.54	0.001	0.0019	0.040	0.0012	0.0013

During the experiment, steel samples were taken during the RH refining process and from the tundish. The specific sampling plan is shown in Table 3. Metallographic specimens (20 mm × 15 mm × 15 mm) were taken from the edges of the steel samples after removing the oxide scales on their surfaces, as shown in Figure 1. After grinding and polishing, the morphology, quantity, size and composition of inclusions with sizes >1 μm in the steel were analyzed by automated SEM/EDS inclusion analysis (ASPEX, FEI, USA), scanning electron microscopy (SEM, JSM-6490LV, JEOL Ltd., Tokyo, Japan), and energy dispersive spectroscopy (EDS, Oxford Instruments, High Wycombe, UK). The contents of total oxygen (T.O.) and nitrogen in the steel were detected by inert gas fusion pulse-infrared absorption spectroscopy.

Table 3. Sampling plan.

No.	Process	Sampling Location	Sampling Method
1	RH refining	Rare earth alloying	Cylindrical sample (Φ70 mm × 100 mm)
2		Carrying out station	Cylindrical sample (Φ70 mm × 100 mm)
3	Tundish	Impact zone	Cylindrical sample (Φ70 mm × 100 mm)
4		Pouring zone	Cylindrical sample (Φ70 mm × 100 mm)

Φ represents the diameter of cylindrical sample.

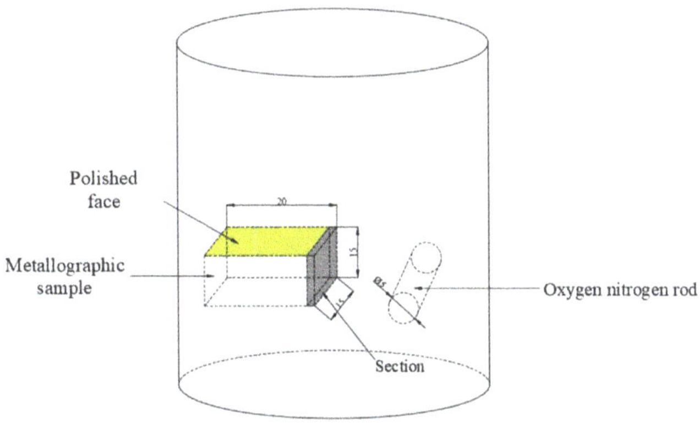

Figure 1. Diagram depicting sampling of the experimental steel.

The magnetic flux density, B_{50}, was determined at a magnetic field strength of H = 5000 A/m, and the core losses, $P_{15/50}$, at an induction of 1.5 T and 50 Hz. Magnetic measurements were carried out for final-annealed sheets of 30 mm in width and 100 mm in length, in both rolling and transverse directions. The measured values were averaged to align with the Epstein method, using MPG100D equipment (Dr. Brockhaus Messtechnik GmbH & Co. KG, Lüdenscheid, Germany).

3. Results

The morphologies and compositions of inclusions in the steel were analyzed by SEM and EDS. Figures 2 and 3 show the typical morphologies, compositions and surface scanning results for inclusions in the steel during the RH refining process, with rare earth treatment. The main types of inclusions in the steel are (La, Ce)Al_2O_3+CaO-Al_2O_3-MgO,

(La, Ce)Al_2O_3+CaS-(CaO-Al_2O_3), and CaO-Al_2O_3-MgO. Inclusions with (La, Ce)Al_2O_3 as the core are surrounded by a CaO-Al_2O_3-MgO or CaO-Al_2O_3-MgO composite phase, and their morphologies are spherical or ellipsoidal. Among the inclusions, the (La, Ce)Al_2O_3 inclusions are distributed in light, white blocks, and CaO-Al_2O_3-MgO inclusions are spherical or approximately spherical.

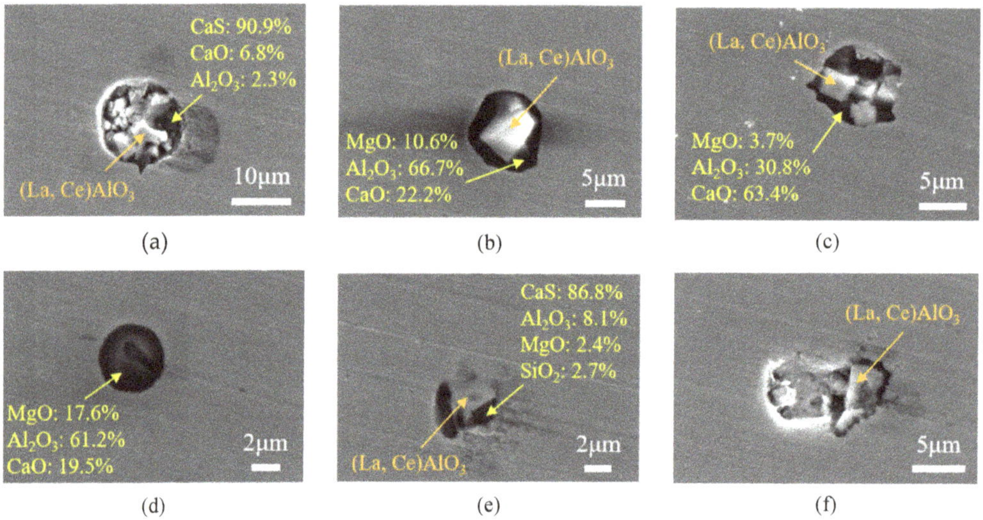

Figure 2. Typical morphologies and compositions of inclusions, with rare earth treatment. (**a**) (La,Ce)AlO_3+Al_2O_3-CaO-CaS; (**b**) Inner layer: (La,Ce)AlO_3, Outer layer:Al_2O_3-CaO-MgO; (**c**) (La,Ce)AlO_3+Al_2O_3-CaO-MgO; (**d**) Al_2O_3-CaO-MgO; (**e**) (La,Ce)AlO_3+Al_2O_3-MgO-CaS; (**f**) (La,Ce)AlO_3.

Figures 4 and 5 show the typical morphologies, compositions and surface scanning results for inclusions in the steel at the end of the RH refining process. It was found that the types of inclusions are still (La, Ce)Al_2O_3+CaO-Al_2O_3-MgO, (La, Ce)Al_2O_3+CaS+oxides, and CaO-Al_2O_3-MgO. (La, Ce)Al_2O_3 inclusions are distributed in long strips at the edge of the composite phase. In addition, some inclusions contain a small amount of SiO_2.

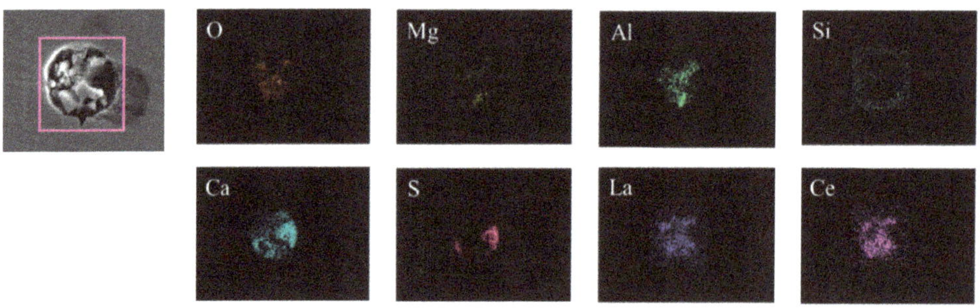

(**a**)

Figure 3. *Cont.*

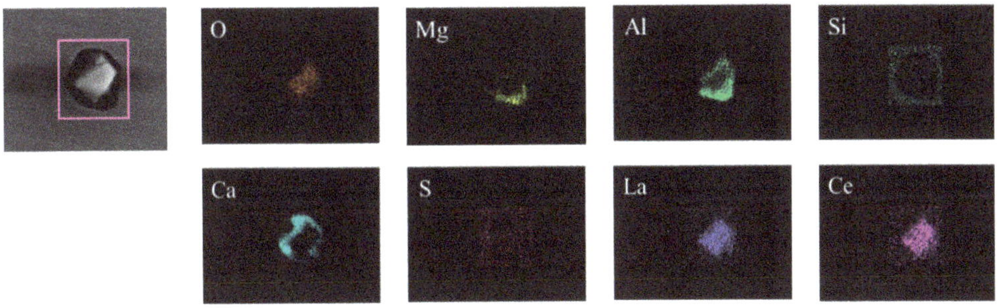

(**b**)

Figure 3. Surface scanning results for inclusions, with rare earth treatment. (**a**) (La, Ce)Al_2O_3+CaS-CaO-Al_2O_3, (**b**) (La, Ce)Al_2O_3+CaO-Al_2O_3-MgO.

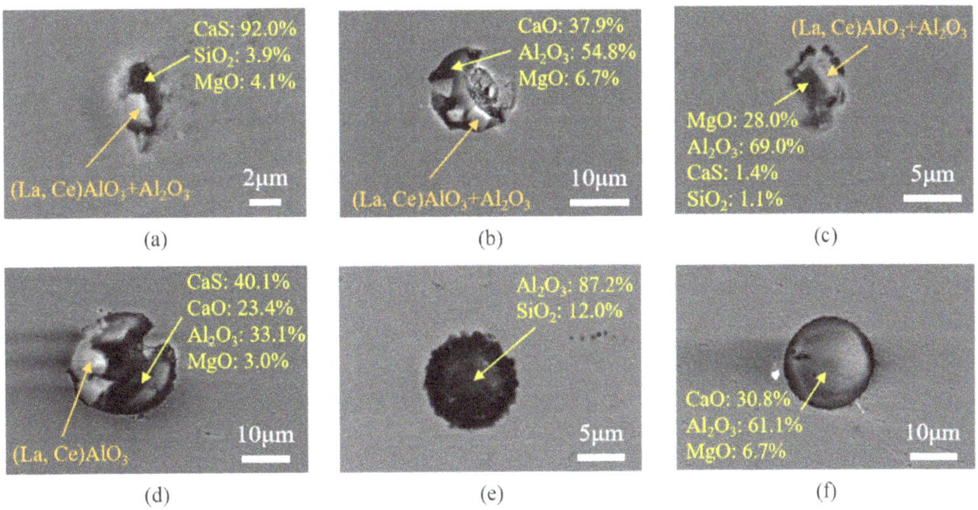

Figure 4. Typical morphologies and compositions of inclusions in the steel at the end of the RH refining process. (**a**) (La,Ce)AlO_3+Al_2O_3+MgO-CaS-SiO_2; (**b**) (La,Ce)AlO_3+Al_2O_3+Al_2O_3-CaO-MgO; (**c**) (La,Ce)AlO_3+Al_2O_3+Al_2O_3-MgO-CaS-SiO_2; (**d**) (La,Ce)AlO_3+Al_2O_3-CaO-MgO-CaS; (**e**) Al_2O_3-SiO_2; (**f**) Al_2O_3-CaO-MgO.

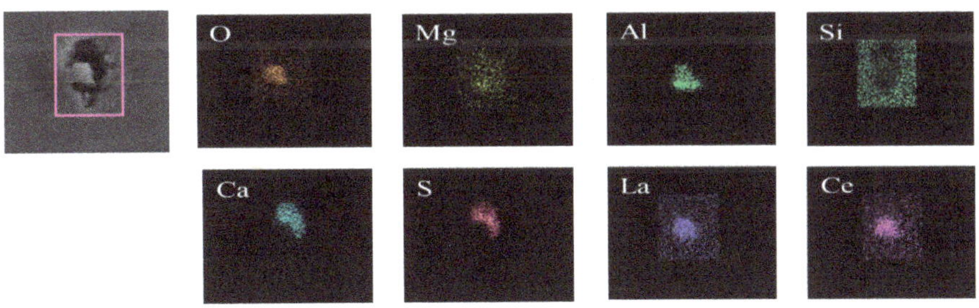

(**a**)

Figure 5. *Cont.*

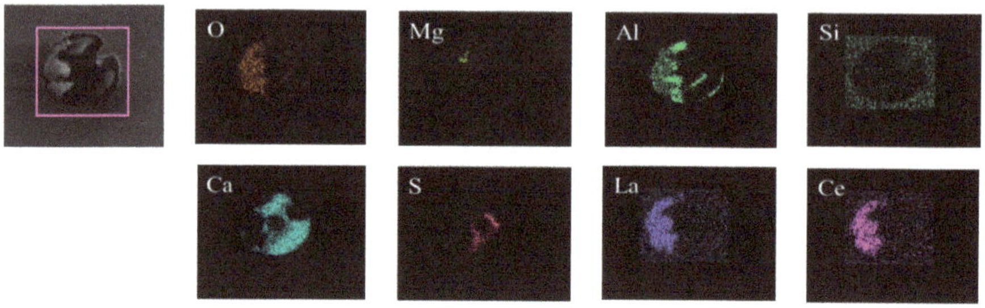

(**b**)

Figure 5. Surface scanning results for inclusions in the steel at the end of the RH refining process. (**a**) (La, Ce)Al_2O_3+CaS-MgO-Al_2O_3, (**b**) (La, Ce)Al_2O_3+CaS-MgO-Al_2O_3-CaO.

The main types of inclusions in the tundish impact zone are (La, Ce)Al_2O_3+CaO-Al_2O_3-MgO, (La,Ce)Al_2O_3+CaS+oxides and CaO-Al_2O_3-MgO-(CaS); the typical morphologies and compositions are shown in Figure 6. Compared to the inclusions found after the RH refining process, the morphologies of inclusions in the tundish impact zone are mainly spherical. Figure 7 shows surface scanning results for typical inclusions. The (La, Ce)Al_2O_3+CaO-Al_2O_3-MgO inclusions are composed of two composite phases with (La, Ce)Al_2O_3 as the core, surrounded by CaO-Al_2O_3-MgO.

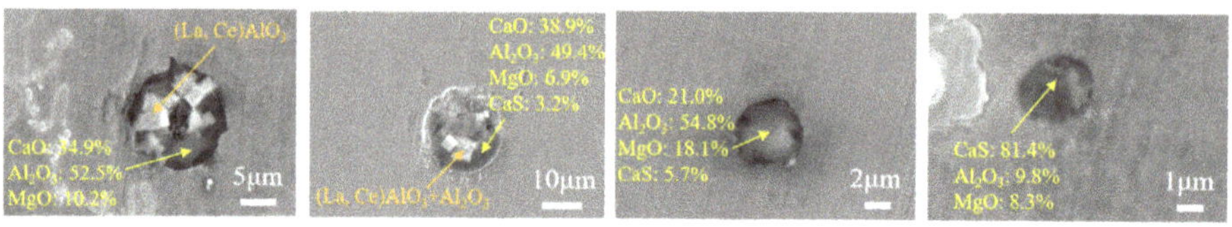

(**a**) (**b**) (**c**) (**d**)

Figure 6. Typical morphologies and compositions of inclusions in the tundish impact zone. (**a**) (La,Ce)AlO_3+Al_2O_3-MgO-CaO; (**b**) (La,Ce)AlO_3-Al_2O_3+Al_2O_3-CaO-MgO-CaS; (**c**) Al_2O_3-MgO-CaS-CaO; (**d**) Al_2O_3-MgO-CaS.

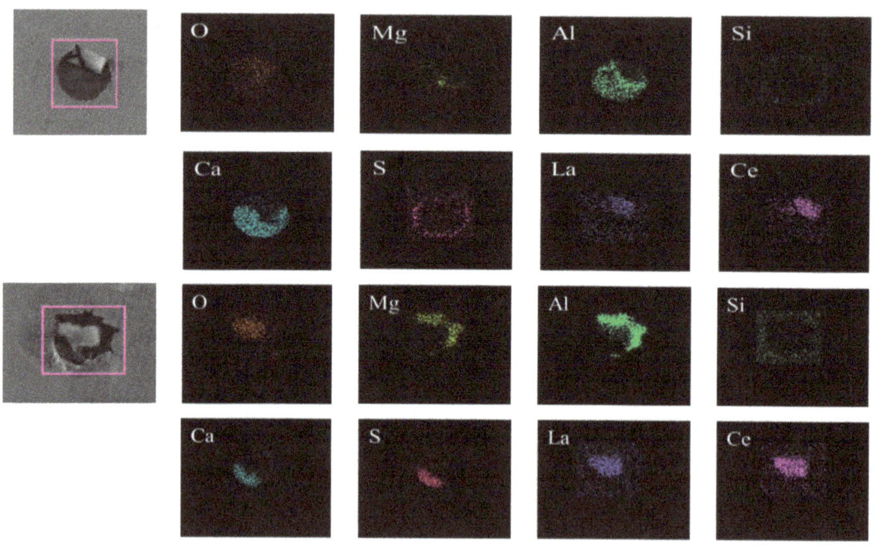

Figure 7. Surface scanning results for (La, Ce)Al_2O_3+CaO-Al_2O_3-MgO in the tundish impact zone.

Figures 8 and 9 show the typical morphologies, compositions and surface scanning results for inclusions in the tundish pouring zone. The types of inclusions in the tundish pouring zone are mainly (La, Ce)Al_2O_3+CaO-Al_2O_3-MgO, (La, Ce)Al_2O_3+CaS+oxides, and MgO-Al_2O_3-SiO_2-CaS. The (La, Ce)Al_2O_3 inclusions are distributed in light, white blocks.

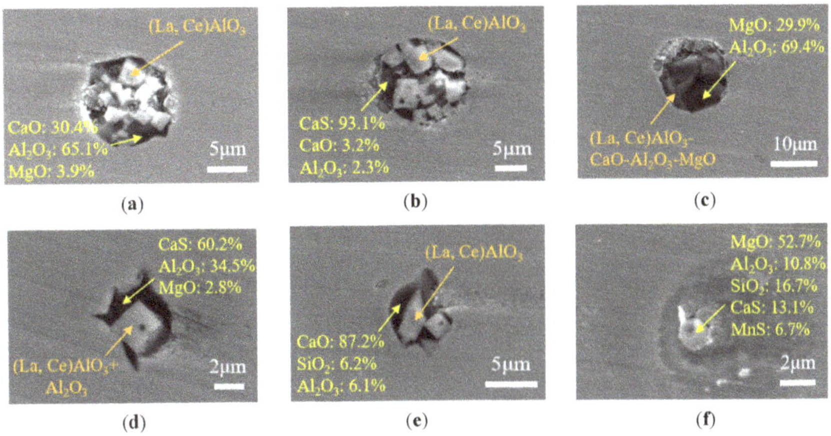

Figure 8. Typical morphologies and compositions of inclusions in the tundish pouring zone. (**a**) (La,Ce)AlO_3+Al_2O_3-MgO-CaO; (**b**) (La,Ce)AlO_3+Al_2O_3-CaO-CaS; (**c**) (La,Ce)AlO_3-Al_2O_3-MgO-CaO+Al_2O_3-MgO; (**d**) (La,Ce)AlO_3-Al_2O_3+Al_2O_3-MgO-CaS; (**e**) (La,Ce)$AlO_{3+}$$Al_2O_3$-CaO-$SiO_2$; (**f**) Al_2O_3-CaS-MgO-SiO_2-MnS.

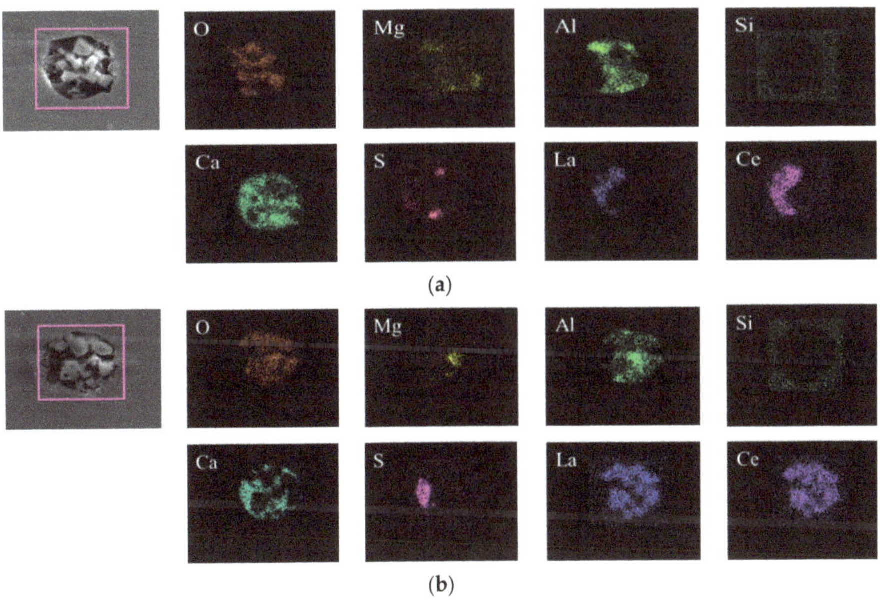

Figure 9. Surface scanning results for inclusions in the tundish pouring zone. (**a**) (La, Ce)Al_2O_3+MgO-Al_2O_3-CaO, (**b**) (La, Ce)Al_2O_3+CaS-Al_2O_3-CaO.

4. Discussion

4.1. Changes to Inclusion Types after Rare Earth Treatment

Figure 10 shows the morphologies and compositions of typical inclusions after alloying in the RH refining process, with and without rare earth treatment. It can be seen that without rare earth treatment, the types of inclusions in the steel after RH alloying are mainly Al_2O_3-SiO_2-(MnO) and Al_2O_3-(MgO), and some inclusions contain a small amount of CaS and MnS. The Al_2O_3-(MgO) inclusions are mostly spherical, while those including CaS and MnS have square or angular shapes. In addition, some Al_2O_3 inclusions aggregate in clusters. After adding rare earth La–Ce alloys during the RH refining process, the main types of inclusions are (La, Ce)Al_2O_3+CaO-Al_2O_3-MgO, (La, Ce)Al_2O_3+CaS-(CaO-Al_2O_3), and CaO-Al_2O_3-MgO, a finding which is similar to that of Ren et al. [23,24]. Inclusions with (La, Ce)Al_2O_3 as the core are surrounded by a CaO-Al_2O_3-MgO or CaO-Al_2O_3-MgO composite phase, and the morphologies are spherical or ellipsoidal.

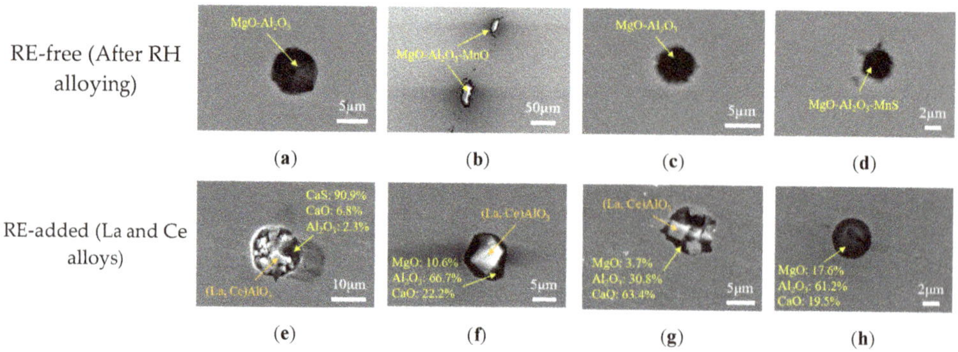

Figure 10. Morphologies and compositions of inclusions, before and after rare earth treatment, after RH alloying. (**a**) Al_2O_3-MgO; (**b**)MgO-Al_2O_3-MnO; (**c**) Al_2O_3-MgO; (**d**) Al_2O_3-MgO-MnS; (**e**) (La,Ce)$AlO_{3+}Al_2O_3$-CaO-CaS; (**f**) (La,Ce)$AlO_{3+}Al_2O_3$-MgO-CaO; (**g**) (La,Ce)$AlO_{3+}Al_2O_3$-MgO-CaO; (**h**) Al_2O_3-MgO-CaO.

Figure 11 shows the morphologies and compositions of typical inclusions in the steel before and after rare earth treatment, after the RH refining station. The majority of inclusions at the RH station are composed of Al_2O_3-CaO, with some inclusions containing a small amount of MnS and CaS. The content of Al_2O_3 in the inclusions is high, and their morphologies are spherical or irregular. After rare earth treatment, the main types of inclusions in the steel after the RH refining station are (La, Ce)Al_2O_3+CaO-Al_2O_3-MgO, (La, Ce)Al_2O_3+CaS+oxides, and CaO-Al_2O_3-MgO. The (La, Ce)Al_2O_3 inclusions are distributed at the edge of the composite phase. In addition, some of the inclusions contain a small amount of SiO_2.

Figure 12 shows the typical morphologies and compositions of inclusions in the tundish pouring zone, before and after rare earth treatment. The results show that without rare earth treatment, the inclusions in the tundish pouring zone are composed of Al_2O_3-CaO-MgO-(SiO_2), and their morphologies are spherical or oval. After adding rare earth, the main types of inclusions in the steel are (La, Ce)Al_2O_3+CaO-Al_2O_3-MgO, (La, Ce)Al_2O_3+CaS+oxides, and MgO-Al_2O_3-CaS.

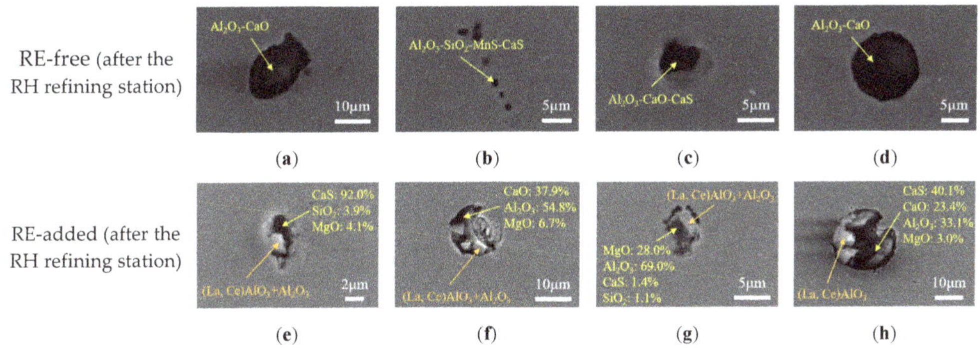

Figure 11. Morphologies and compositions of typical inclusions in the steel, before and after rare earth treatment, after the RH refining station. (**a**) Al_2O_3-CaO; (**b**) Al_2O_3-SiO_2-MnS-CaS; (**c**) Al_2O_3-CaO-CaS; (**d**) Al_2O_3-CaO; (**e**) (La,Ce)$AlO_{3+}Al_2O_3$-MgO-CaS-SiO_2; (**f**) (La,Ce)$AlO_{3-}Al_2O_3$+Al_2O_3-MgO-CaO; (**g**) (La,Ce)AlO_3+Al_2O_3-MgO-CaS-SiO_2; (**h**) (La,Ce)$AlO_{3+}Al_2O_3$-MgO-CaS-CaO.

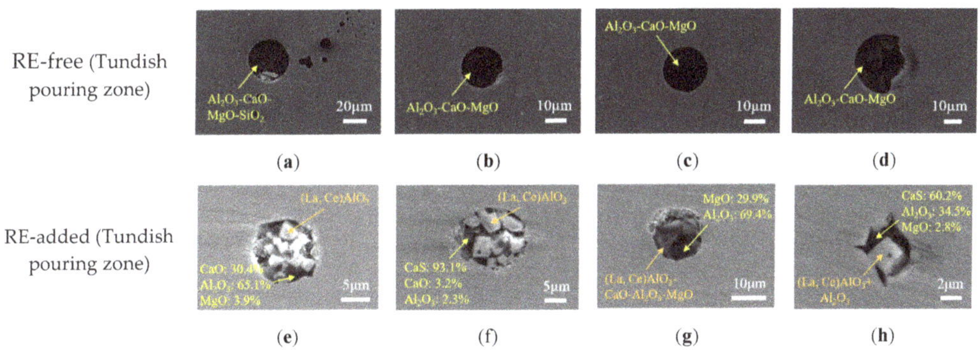

Figure 12. Morphologies and compositions of typical inclusions in the steel, before and after rare earth treatment, in the tundish pouring zone. (**a**) Al_2O_3-CaO-MgO-SiO_2; (**b**) Al_2O_3-CaO-MgO; (**c**) Al_2O_3-CaO MgO; (**d**) Al_2O_3-CaO-MgO; (**e**) (La,Ce)$AlO_{3+}Al_2O_3$-MgO-CaO; (**f**) (La,Ce)AlO_3+Al_2O_3-CaS-CaO; (**g**) (La,Ce)AlO_3+Al_2O_3-MgO; (**h**) (La,Ce)AlO_3-Al_2O_3+Al_2O_3-MgO-CaS.

Through the analysis of the experimental results before and after rare earth treatment in different processes, it can be seen that rare earth treatment transforms the typical CaO-Al_2O_3-MgO inclusions with rectangular, acicular, or irregular shapes, into spherical or ellipsoidal rare earth inclusions. According to the investigations of Li et al. [17], rare earth elements added to liquid steel at a certain temperature will first react with the non-metallic elements O, S, etc., to form inclusions. The general formula of the chemical reaction is as follows:

$$[\mathrm{RE}] + \frac{y}{x}[\mathrm{M}] = \frac{1}{x}\mathrm{RE}_x\mathrm{M}_y(\mathrm{s}) \quad (1)$$

where [RE] represents various rare earth elements dissolved in liquid steel, [M] represents various impurity elements dissolved in liquid steel, and (s) refers to the solid phase, with an activity of 1. According to the change of Gibbs free energy, the different types of inclusions can be determined. In addition to the formation sequence of inclusions in liquid steel, rare earth inclusions can also be converted to each other, which is closely related to the amounts of rare earth elements added. In the previous study, the author introduced the modification of inclusions in W350 non-oriented silicon steel by rare earth Ce [25].

Furthermore, it is generally accepted that rare earth inclusions have high melting points, that is, higher than 1690 °C. The temperature of liquid steel during the RH refining process and the continuous casting process is lower than the melting point of rare earth inclusions. Therefore, the physical properties of rare earth inclusions essentially determine that they can be used as the basic conditions for heterogeneous nucleation in liquid steel [26]. Whether during the RH refining process or the tundish process, the types of inclusions in the steel changed significantly before and after rare earth treatment. With the ability of rare earth oxides (sulfides) to form heterogeneous nuclei, the types of inclusions with rare earth treatment are mainly rare earth composite inclusions, and most of them are (La, Ce)Al_2O_3+conventional inclusions [27]. In addition, the inclusions detected after rare earth treatment are usually composed of more than two composite phases, indicating that the addition of rare earth elements promotes the agglomeration of inclusions.

4.2. Size Distribution of Inclusions before and after Rare Earth Treatment

Figure 13 shows the size distributions of inclusions, the proportions of inclusions with different sizes, and the changes in average size during the production processes of RH refining→after RH station→tundish. The morphologies, quantities and sizes of inclusions with a size of >1 μm in the steel were analyzed by an automatic scanning electron microscope (ASPEX). It can be seen that the number of inclusions with a size of 1.0~3.5 μm in the steel during the RH refining process, after rare earth treatment, is increased and their proportion is 11.11~24.07%. The average size of inclusions in the steel is 2.66 μm. The number of inclusions in the steel increases after the RH station, and the number of inclusions larger than 5 μm in the steel increases due to the collision of inclusions during the RH refining process. As the pouring casting proceeds, the number of micro inclusions with sizes of 1.0~2.5 μm in the steel decreases, while the number of inclusions with sizes of >5 μm increases, especially in the tundish impact zone. The number density of inclusions in the tundish pouring zone is 8.50/mm^2 and their proportion is 28.57%. The size distribution of inclusions is relatively stable.

Inclusions of less than 1 μm were difficult to observe due to their small size. Therefore, high-magnification TEM (backscattered, JEM-2100, JEOL Ltd, Japan) observation was used to assist in the identification of typical inclusions, through energy spectrum observation statistics.

The carbon extraction replica test samples for TEM were prepared as samples with sizes of 8 mm (TD) × 10 mm (RD) by wire cutting, and then roughly and finely ground. The samples were prepared by electro-polishing at 90 mA in 10% AA (acid alcohol) electrolyte for 120 s. The electrolyzed samples were coated with a layer of carbon film with a thickness of approximately 30 nm, using a vacuum carbon spray instrument. After dividing the carbon film into sizes of approximately 2 mm × 2 mm, they were placed in a 10% perchloric acid alcohol solution for electrolytic release, and then molybdenum nets with a 3 mm diameter were used to extract the carbon film. The samples were also prepared for TEM, and then electropolishing and observations, with 30 fields in each sample, were performed at different magnifications under TEM. The sizes and numbers of precipitates were analyzed using Image-Pro Plus.

The inclusions in hot-rolled sheet were observed and counted, and 30 fields of view were randomly selected, as shown in Figure 14.

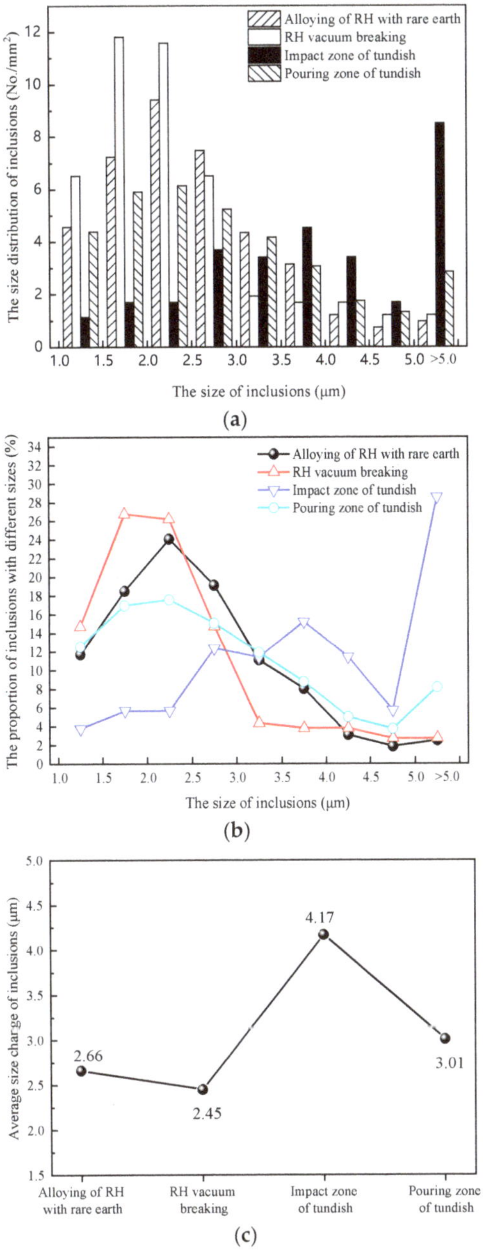

Figure 13. Size distributions and changes to inclusions in the steel. (**a**) The size distribution of inclusions, (**b**) The proportion of inclusions with different sizes, (**c**) Average size change of inclusions.

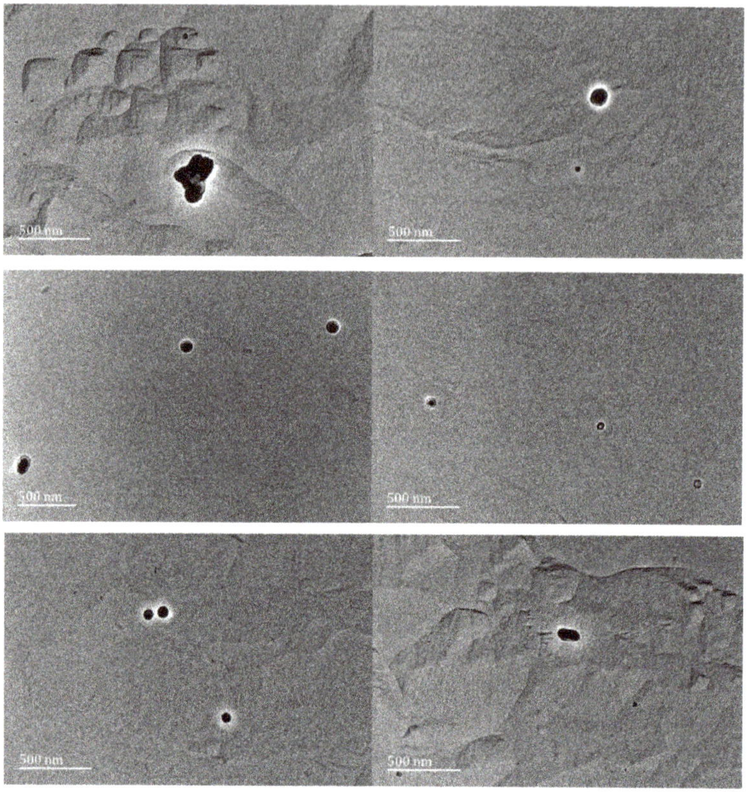

Figure 14. TEM morphology of micro inclusions in hot-rolled sheet, after rare earth treatment.

As illustrated in Figure 15, the content of micro inclusions (≤1 μm) in hot-rolled sheets is 51,458.2/mm^2 and their average size is 0.388 μm, before rare earth treatment. After rare earth treatment, the content of micro inclusions (≤1 μm) in hot-rolled sheets is 24,230.2/mm^2 and their average size is 0.427 μm. According to previous investigations, when the size of micro inclusions in non-oriented silicon steel is 0.1~0.5 μm, the magnetic properties of non-oriented silicon steel seriously deteriorate because the sizes of the micro inclusions are close to the thickness of the magnetic domain wall. In the current study, the number of micro inclusions in hot-rolled sheets after rare earth treatment is greatly reduced, and the average size of the micro inclusions is also increased. The magnetic properties of the steel with rare earth treatment are improved, which is consistent with previous investigations [28–30]. Compared to sheet produced through the original process, the iron loss of the 0.35 mm finished experimental sheet is reduced by 0.068 W/kg, and the magnetic induction is increased by 0.007 T, as shown in Table 4. The iron loss of the 0.50 mm finished experimental sheet is reduced by 0.008 W/kg, and the magnetic induction is increased by 0.004 T.

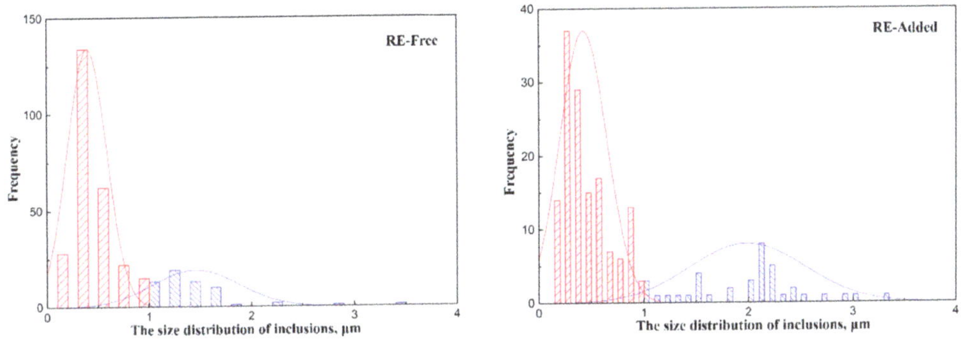

(a) (b)

Figure 15. Size distribution of inclusions before and after rare earth treatment. (**a**) RE-free (**b**) RE-added.

Table 4. Electromagnetic properties of the steel.

Grade	Project	$P_{1.5/50}$, W/kg	Δ Value ($P_{1.5/50}$, W/kg)	B_{50}, T	Δ Value, (B_{50}, T)
0.35 mm	Experiment roll after rare earth treatment	2.370	−0.068	1.679	+0.007
	Original process sheet	2.438		1.672	
0.50 mm	Experiment roll after rare earth treatment	2.730	−0.008	1.695	+0.004
	Original process sheet	2.738		1.691	

5. Conclusions

(1) The types of inclusions in the steel significantly change after rare earth treatment. They are mainly composed of (La, Ce)Al_2O_3 and conventional inclusions. The morphologies are mostly blocky and partially chain-shaped.

(2) Adding rare earth La–Ce alloys during the RH refining process results in inclusions that are mainly (La, Ce)Al_2O_3+CaO-Al_2O_3-MgO, (La, Ce)Al_2O_3+CaS+oxides, and CaO-Al_2O_3-MgO. At the end of RH refining, the inclusions are mainly (La, Ce)Al_2O_3+CaO-Al_2O_3-MgO, (La,Ce)Al_2O_3+CaS+oxides, and CaO-Al_2O_3-MgO. The inclusions in the tundish impact zone are mainly (La, Ce)Al_2O_3+CaO-Al_2O_3-MgO, (La, Ce)Al_2O_3+CaS+oxides, and CaO-Al_2O_3-MgO (CaS). The inclusions in the tundish pouring zone are mainly (La, Ce)Al_2O_3+CaO-Al_2O_3-MgO, (La, Ce)Al_2O_3+CaS+oxides, MgO-Al_2O_3-SiO_2-CaS, and Al_2O_3-SiO_2-CaO. After rare earth treatment during the RH refining process, the number of inclusions in the steel with sizes of 1.0~3.5 μm is increased and their average size is 2.66 μm. Due to collisions and growth, the number of inclusions with sizes greater than 4 μm in the steel increases after the RH refining station. The number of micro inclusions with sizes of 1.0~2.5 μm in the tundish decreases, while the number of inclusions with sizes larger than 5 μm increases.

(3) Before rare earth treatment, the content of micro inclusions (≤1 μm) in hot-rolled sheets is 51,458.2/mm^2 and their average size is 0.388 μm. The content of micro inclusions (≤1 μm) in hot-rolled sheets with rare earth treatment is 24,230.2/mm^2 and their average size is 0.427 μm. Compared to sheets produced by the original process, the magnetic properties of the experimental sheets show that the iron loss of the 0.35 mm finished product is decreased by 0.068 W/kg, and the magnetic induction is increased by 0.007 T. The iron loss of the 0.5 mm finished product is decreased by 0.008 W/kg, and the magnetic induction is increased by 0.004 T.

Author Contributions: Conceptualization, H.W. and H.L.; methodology, H.W.; software, H.W.; validation, H.W., Y.Z. and W.Z.; formal analysis, H.W.; investigation, H.W.; resources, H.W.; data curation, H.W., J.Q. and Y.N.; writing—original draft preparation, H.W.; writing—review and editing, H.W. and S.Q.; visualization, H.W. and H.L.; supervision, S.Q.; project administration, H.W. and S.Q.; funding acquisition, H.W. All authors have read and agreed to the published version of the manuscript.

Funding: This research was funded by the National Natural Science Foundation of China (Nos. 52274312 and 51804003), the Fund of Education Department of Anhui Province (Nos. 2022AH050291, 2022AH050293), the Open Project Program of Anhui Province Key Laboratory of Metallurgical Engineering & Resources Recycling (Anhui University of Technology) (Nos. SKF21-04 and SKF21-05), the University Natural Science Research Project of Education Department of Anhui Province (No. KJ2021A0396), and the Jiangxi Province Major Scientific and Technological Research and Development Special Funding Project (20213AAE01009).

Institutional Review Board Statement: Not applicable.

Informed Consent Statement: Not applicable.

Data Availability Statement: Data available in a publicly accessible repository.

Conflicts of Interest: The authors declare no conflict of interest.

References

1. Kubota, T. Recent progress on non-oriented silicon steel. *Steel Res. Int.* **2005**, *76*, 464–470. [CrossRef]
2. Oda, Y.; Kohno, M.; Honda, A. Recent development of non-oriented electrical steel sheet for automobile electrical devices. *J. Mag. Mag. Mater.* **2008**, *320*, 2430–2435. [CrossRef]
3. Paolinelli, S.C.; Da Cunha, M.A. Development of a new generation of high permeability non-oriented silicon steels. *J. Mag. Mag. Mater.* **2006**, *304*, e596–e598. [CrossRef]
4. Kubota, T. Development of non-oriented electrical steel used for high efficiency cores. *Curr. Adv. Mater. Process.* **2003**, *16*, 626.
5. Qin, J.; Yang, P.; Mao, W.M.; Ye, F. Effect of texture and grain size on the magnetic flux density and core loss of cold-rolled high silicon steel sheets. *J. Mag. Mag. Mater.* **2015**, *393*, 537–543. [CrossRef]
6. Oda, Y.; Okubo, T.; Takata, M. Recent development of non-oriented electrical steel in JFE steel. *JFE Tech. Rep.* **2016**, *21*, 7–13.
7. Manoh, P.A.; Ferry, M.; Chandra, T. Five decades of the Zener equation. *ISIJ Int.* **1998**, *38*, 913–924. [CrossRef]
8. Wang, J.Y.; Ren, Q.; Luo, Y.; Zhang, L.F. Effect of non-metallic precipitates and grain size on core loss of non-oriented electrical silicon steels. *J. Mag. Mag. Mater.* **2018**, *451*, 454–462. [CrossRef]
9. Hou, C.K.; Liao, C.C. Effect of cerium content on the magnetic properties of non-oriented electrical steels. *ISIJ Int.* **2008**, *48*, 531–539. [CrossRef]
10. Ren, Q.; Hu, Z.Y.; Cheng, L.; Zhang, L.F. Effect of rare earth elements on magnetic properties of non-oriented electrical steels. *J. Mag. Mag. Mater.* **2022**, *560*, 169624. [CrossRef]
11. Fan, L.F.; Zhu, R.; He, J.Z.; Lu, B. Effect of rare earth element La on texture and inclusion of non-oriented electrical steel produced by thin slab casting and rolling process. *ISIJ Int.* **2018**, *58*, 2348–2353. [CrossRef]
12. Zhang, F.; Ma, C.S.; Wang, B.; Zhang, P.L.; Ma, Z.G.; Zhang, Y. Control of nonmetallic inclusions of non-oriented silicon steel sheets by the rare earth treatment. *Baosteel Tech. Res.* **2011**, *5*, 41–45.
13. He, Z.H.; Sha, Y.H.; Gao, Y.K.; Chang, S.T.; Zhang, F.; Zuo, L. Recrystallization texture development in rare-earth (RE)-doped non-oriented silicon steel. *J. Iron Steel Res. Int.* **2020**, *27*, 1339–1346. [CrossRef]
14. Shi, C.J.; Jin, Z.L.; Ren, H.P.; You, J.L. Effect of lanthanum on recrystallization behavior of non-oriented silicon steel. *J. Rare Earths* **2017**, *35*, 309–314. [CrossRef]
15. Takashima, M.; Ono, T.; Nishimura, K. Low iron loss non-oriented electrical steel for high efficiency motors "RMA Series". *Kawasaki Steel Tech. Rep.* **1998**, *39*, 48–49.
16. Takashima, M.; Morito, N.; Honda, A.; Maeda, C. Non-oriented electrical steel sheet with low iron loss for high-efficiency motor cores. *IEEE Trans. Mag.* **1999**, *35*, 557–561. [CrossRef]
17. Li, N.; Lu, Q.Y.; Wang, Y.Q.; Zhu, Z.H.; Xiang, L.X.; Qiu, S.T. Effect of Ce on inclusions modification in 2.9% Si-0.8% A1 non-oriented electrical steel. *J. Iron Steel Res.* **2017**, *29*, 570–576.
18. Kong, W.; Chen, Y.; Cang, D. A statistical study of inclusions in medium-grade non-oriented silicon steel. *Metall. Res. Technol.* **2019**, *116*, 207–214. [CrossRef]
19. Qiao, J.L.; Guo, F.H.; Hu, J.W.; Xiang, L.; Qiu, S.T.; Wang, H.J. Precipitates in compact strip production (CSP) process non-oriented electrical steel. *Metals* **2020**, *10*, 1301. [CrossRef]
20. Wan, Y.; Chen, W.Q.; Wu, S.J. Effect of lanthanum content on microstructure and magnetic properties of non-oriented electrical steels. *J. Rare Earths* **2013**, *3*, 727–733. [CrossRef]
21. Wan, Y.; Wu, S.J.; Li, J. Effects of complex trace elements on the microstructure and magnetic properties of non-oriented electrical steels. *Metall. Res. Technol.* **2016**, *113*, 101–109. [CrossRef]

22. Wan, Y.; Chen, W.Q.; Wu, S.J. Effects of lanthanum and boron on the microstructure and magnetic properties of non-oriented electrical steels. *High Temp. Mater. Process.* **2014**, *33*, 115–121. [CrossRef]
23. Ren, Q.; Zhang, L.F. Effect of cerium content on inclusions in an ultra-low-carbon aluminum-killed steel. *Metall. Mater. Trans. B* **2020**, *51*, 589–600. [CrossRef]
24. Ren, Q.; Zhang, L.F.; Hu, Z.Y.; Cheng, L. Transient influence of cerium on inclusions in an Al-killed non-oriented electrical steel. *Ironmak. Steelmak.* **2020**, *48*, 191–199. [CrossRef]
25. Wang, H.J.; Niu, Y.H.; Ling, H.T.; Qiao, J.L.; Zhang, Y.L.; Zhong, W.; Qiu, S.T. Modification of rare earth Ce on inclusions in W350 non-oriented silicon steel. *Metals* **2023**, *13*, 453. [CrossRef]
26. Ji, Y.P.; Li, Y.M.; Zhang, M.X.; Qu, W.; Zhao, T.X.; Ren, H.P. Grain refinement mechanism of the δ-ferrite in steels through cerium addition. *Metall. Mater. Trans. A* **2020**, *51*, 1707–1718. [CrossRef]
27. Li, B.; Zhu, H.Y.; Zhao, J.X.; Song, M.M.; Li, J.L.; Xue, Z.L. Effect of rare-earth La on inclusion evolution in high-Al steel. *Steel Res. Int.* **2022**, *93*, 2100347. [CrossRef]
28. Bóc, I.; Cziráki, Á.; Gróf, T.; Csébi, J. Analysis of inclusions in cold-rolled n.o. Si-Fe strips. *J. Mag. Mag. Mater.* **1990**, *83*, 381–383. [CrossRef]
29. Zhang, F.; Li, G.Q.; Miao, L.D.; Zhu, C.Y. Effects of chemical composition systems on non-metallic inclusions in non-oriented silicon steel sheets. *J. Iron Steel Res.* **2012**, *24*, 40–44.
30. Chen, L.F.; Zhao, Z.Y. Effect of cooling rate on precipitating behavior of inclusions in non-oriented silicon steel. *Spec. Steel* **2015**, *36*, 55–58.

metals MDPI

Article

Evolution of Inclusions and Cleanliness in Ti-Bearing IF Steel Produced via the BOF–LF–RH–CC Process

Baohui Yuan [1], Jianhua Liu [1,*], Jianhua Zeng [2], Min Zhang [2], Jihong Huang [3] and Xiaodong Yang [1,3]

1 National Engineering Research Center for Advanced Rolling and Intelligent Manufacturing, University of Science and Technology Beijing, Beijing 100083, China; yuanbaohui20@163.com (B.Y.); xiaodong2984@163.com (X.Y.)
2 Pangang Group Research Institute Co., Ltd., Panzhihua 617000, China; yjyzjh@126.com (J.Z.); muyufile@163.com (M.Z.)
3 Pangang Group Xichang Steel and Vanadium Co., Ltd., Xichang 615032, China; huangjihong2022@163.com
* Correspondence: liujianhua@metall.ustb.edu.cn

Abstract: Owing to the insufficient converter heat, IF steel is produced via the BOF–LF–RH–CC process in Pangang Group Xichang Steel and Vanadium Co., Ltd. To clarify the evolution of inclusions and the control strategy to improve the cleanliness of molten steel in Ti-bearing IF steel produced via the long process, scanning electron microscopy with energy spectroscopy analysis and automatic scanning electron microscopy were employed to analyze the number, size, type and morphology of inclusions in IF steel from RH to tundish. The results show that the characteristics of inclusions are similar in two heats during RH treatment. In the tundish sample of Heat 2, the number density (ND) and area fraction (AF) of Al_2O_3 and $Al_2O_3 \cdot TiO_x$ inclusions increase significantly, and the size of Al_2O_3 inclusions decreases obviously, which is closely related to the serious reoxidation of molten steel caused by the slag with high oxidability during the holding process. Meanwhile, a new method of determining the number of cluster inclusions is used to evaluate the cleanliness of IF steel in this paper, and the obtained number of inclusion clusters is consistent with the trend of ND and AF of inclusions. The effects of reoxidation on the morphology, number and other indexes of Al_2O_3 and $Al_2O_3 \cdot TiO_x$ inclusions are discussed in detail, and there are two ways of forming $Al_2O_3 \cdot TiO_x$ inclusions in the case of serious reoxidation. To weaken the reoxidation process and enhance the cleanliness of IF steel produced via the long process, reducing the oxygen content in molten steel before Al deoxidation, minimizing the holding time and reducing the oxidability of slag after RH are helpful.

Keywords: nonmetallic inclusion; cleanliness; Ti-bearing IF steel; reoxidation; long producing process

Citation: Yuan, B.; Liu, J.; Zeng, J.; Zhang, M.; Huang, J.; Yang, X. Evolution of Inclusions and Cleanliness in Ti-Bearing IF Steel Produced via the BOF–LF–RH–CC Process. *Metals* **2022**, *12*, 434. https://doi.org/10.3390/met12030434

Academic Editor: Mark E. Schlesinger

Received: 27 December 2021
Accepted: 21 February 2022
Published: 1 March 2022

1. Introduction

Ti-bearing IF steel is widely used in automotive, home appliances, electronics and other fields due to its good deep-drawing performance. With the market's increasing requirements for the surface and internal quality of final products, the demand for high-quality IF steel is increasing [1,2]. The performance of IF steel is greatly affected by the type, number and size distribution of nonmetallic inclusions [3–6]. Engine failure caused by nonmetallic inclusions of tens of microns is a typical problem in the automotive industry [7,8]. Clogging of the submerged entry nozzle is also a common problem in the steelmaking process because of the presence of Ti [9–12]. Therefore, it is important to enhance the potential for inclusion removal from molten steel during RH and tundish processes.

There are two routes for producing IF steel currently; one is the BOF–RH–CC short process, and the other is the BOF–LF–RH–CC long process. The short process is more universally employed in this field owing to its simpler process and better refining effect [13–17]. In Pangang Group Xichang Steel and Vanadium Co., Ltd., the hot metal after extracting vanadium with lower carbon content is smelted in BOF, resulting in the insufficient heat of molten steel after BOF. In order to produce smoothly, electric heating (LF) is used to compensate for the heat of molten steel. Therefore, the Ti-bearing IF steel is produced via the BOF–LF–RH–CC process.

At this plant, a high oxygen activity generally in the range of 500~900 ppm after BOF is necessary to complete the deep decarburization task by the natural decarburization process during RH treatment. Due to there being no pre-deoxidation operation before RH and the longer reaction time between the molten steel and slag, the total mass fraction of the FeO and MnO in the ladle slag is basically over 15% during RH treatment. Some scholars have also studied the cleanliness control of IF steel produced via the long process [15,18,19]. However, there are few reports about the effect of the high oxidized slag on inclusions in IF steel produced via the long process, and the influence of the reoxidation process on the cleanliness of molten steel should not be ignored. In the industrial producing process, the evolution of inclusions in Ti-bearing IF steel produced by the long process also needs to be studied more deeply. Additionally, it is still not clear how to take measures to improve the cleanliness of IF steel produced via the long process.

The main objective of this work is to clarify the evolution of inclusions in the long process from RH to tundish and evaluate the influence of the reoxidation process and the relevant process parameters on the cleanliness of molten steel. Theoretical guidance obtained by these experiments is good for optimizing the long smelting process and improving the cleanliness of Ti-bearing IF steel.

2. Materials and Methods

2.1. Industrial Investigations

At the steel plant of Pangang Group Xichang Steel and Vanadium Co., Ltd., IF steel is typically produced in heat sizes of 200 tons via the basic oxygen furnace (BOF)–ladle furnace (LF)–Ruhatahl–Hausen (RH)–continuous casting (CC) process. The processes mainly include the following: (1) The BOF steelmaking is carried out after extracting vanadium. (2) The temperature of molten steel is increased in LF. (3) After ending the decarburization process, aluminum and sponge titanium are added for alloying. The interval time between the added Al and Ti is about 4 min, and the pure circulation time after Ti addition is usually around 6 min. (4) After the vacuum break, the holding time is usually between 30 and 40 min. The steps are shown in Figure 1.

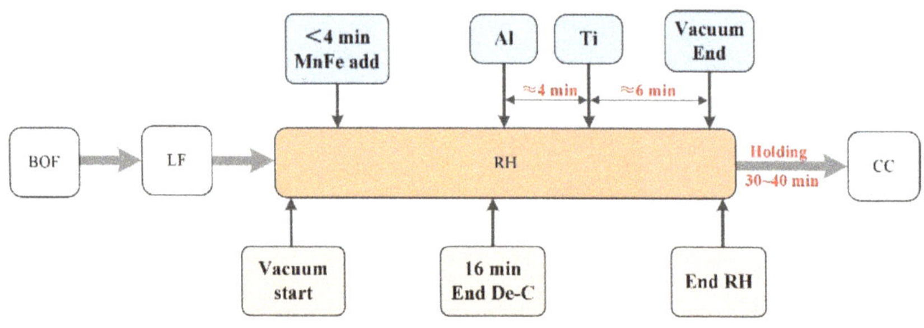

Figure 1. Steps of producing IF steel during steelmaking processes.

The inclusion landscapes of two heats have been tracked from RH to tundish, and 4 standard lollipop samples were taken in different time steps after Al addition. The exact time steps are listed in Table 1. The total oxygen (T.O) and nitrogen (N) contents of samples were analyzed using the TCH-600 oxygen/nitrogen/hydrogen analyzer (LECO Company, San Jose, CA, USA). The chemical compositions of the IF steel slab for two heats are shown in Table 2. The chemical compositions were obtained with ICP-AES (Inductively Coupled Plasma-Atomic Emission Spectrometer).

Table 1. Time of sampling and chemical composition of lollipop samples.

Sample	Heat 1			Heat 2		
	Description	T.O %	N %	Description	T.O %	N %
RH-1	2.0 min after Al addition	0.0080	0.0014	2.5 min after Al addition	0.0120	0.0011
RH-2	3.0 min after Ti addition	0.0027	0.0018	3.3 min after Ti addition	0.0052	0.0010
RH-3	6.0 min after Ti addition	0.0020	0.0018	5.0 min after Ti addition	0.0013	0.0010
Tundish	-	0.0017	0.0020	-	0.0150	0.0014

Table 2. Chemical composition of Ti-bearing IF steel (wt%).

Heats	C	Si	Mn	P	S	Al_s	Ti	N
1	0.0018	0.0030	0.1300	0.0080	0.0060	0.0250	0.0550	0.0027
2	0.0022	0.0030	0.1300	0.0060	0.0060	0.0250	0.0540	0.0018

2.2. Experimental Methods

To analyze the inclusions in the IF steel, the standard samples (15 mm × 15 mm × 15 mm) were cut from the lollipop samples. They were inlaid and ground to 2000 grit and then polished with the diamond polishing agent. The number and size distribution of nonmetallic inclusions (NMIs) were statistically analyzed by an ASPEX Explorer scanning electron microscope (FEI Company, Hillsboro, OR, USA). An area of about 83 mm^2 was set as the standard area on samples for inclusion observation, and the minimum detectable inclusion was 1.0 µm in equivalent circle diameter (ECD). The number density and area fraction are two statistical parameters to characterize the amount of nonmetallic inclusions, as shown in the following two Equations:

$$ND = \frac{n}{A_{total}} \quad (1)$$

$$AF = \frac{A_{inclusion}}{A_{total}} \quad (2)$$

where ND is the number density of inclusions, mm^{-2}; A_{total} is the sample detection area, mm^2; n is the number of detected inclusions on the area of A_{total}; AF is the area fraction of inclusions, 10^{-6}; and $A_{inclusion}$ is the total area of detected inclusions, µm^2.

A Zeiss Sigma 500 field emission scanning electron microscope (ZEISS Company, Oberkochen, Germany) was used to analyze the morphology and composition of inclusions. The inclusions in the view field of metallographic samples were observed and photographed, and energy dispersive spectroscopy (EDS) was used to determine the chemical composition of inclusions by point or plane scanning.

In the present work, the effect of relevant process parameters on the cleanliness of molten steel is further discussed. Besides Heat 1 and Heat 2, several industrial tests were carried out, and the relevant process parameters are listed in Table 3. T.Fe content is the total mass fraction of FeO and MnO in the ladle slag. The ND of inclusions in Table 3 was also analyzed by the ASPEX method and calculated by Equation (1).

Table 3. Parameters of experimental heats during RH and tundish processes.

Heats	[O] before Al Deoxidation/10^{-6}	Holding Time/Min	T.Fe in Slag after RH/%	Number Density of Inclusions/mm^{-2}	
				After RH	Tundish
1	295	35.5	15.3	13.5	9.3
2	294	42.5	18.1	15.2	35.1
3	288	35.3	14.2	9.3	6.1
4	357	26.0	17.0	15.0	2.5
5	247	32.8	16.8	2.5	13.8
6	312	34.9	9.5	9.5	4.6

3. Results and Discussion

3.1. Number and Size Distribution of Inclusions

To reveal the evolution process, the results of the ASPEX inclusion assessment for Heat 1 and Heat 2 were employed to analyze the number and size changes of inclusions from RH to tundish, as shown in Figure 2.

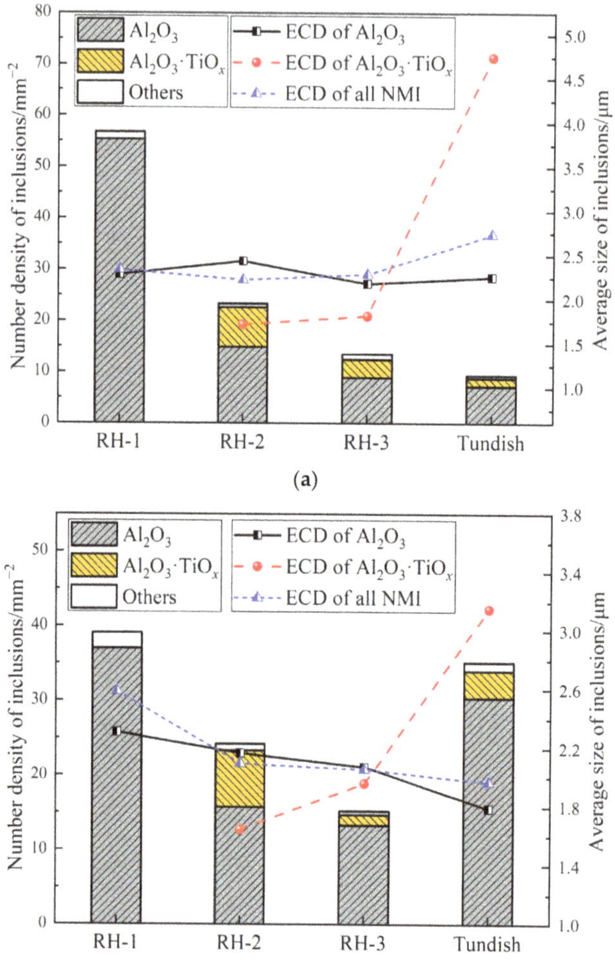

Figure 2. Number and size of different inclusions in (**a**) Heat 1 and (**b**) Heat 2 during RH treatment and tundish processes.

It can be seen from Figure 2 that inclusions in molten steel are mainly divided into three types, namely Al_2O_3, $Al_2O_3 \cdot TiO_x$ and other inclusions. Al_2O_3 and $Al_2O_3 \cdot TiO_x$ inclusions account for almost 100% during the whole steelmaking process. The ND of Al_2O_3 inclusions in molten steel is over 35 mm^{-2} within 3 min after Al addition, and the average size of Al_2O_3 inclusions is about 2.3 μm. The number of Al_2O_3 inclusions decreases significantly at about 3 min after Ti addition and then further decreases at the end of RH. During the whole RH refining process, the size of Al_2O_3 inclusions changes a little. After adding the sponge titanium, the formation of many small $Al_2O_3 \cdot TiO_x$ inclusions causes the reduced size of all NMIs, and the size of $Al_2O_3 \cdot TiO_x$ inclusions in the pure circulation process shows a growing trend. In the tundish sample of Heat 1, the mean size of all inclusions increases to about 2.7 μm owing to the notable increasing size of $Al_2O_3 \cdot TiO_x$ inclusions. On the contrary to Heat 1, the number of various inclusions in the tundish sample of Heat 2 has an increasing trend. In particular, the number of Al_2O_3 inclusions in tundish is more than twice as much as that at the end of RH. The mean size of all inclusions shows a decreasing trend, which is also related to the increase in small Al_2O_3 inclusions.

Figure 3 shows the change of the AF of inclusions during RH treatment and tundish processes. In Figure 3, the AF of inclusions shows basically the same trend with the ND of inclusions. In the tundish sample of Heat 2, the AF of Al_2O_3 and $Al_2O_3 \cdot TiO_x$ inclusions also increases significantly. The process characteristics for producing IF steel in this steel plant are the main reason for this phenomenon.

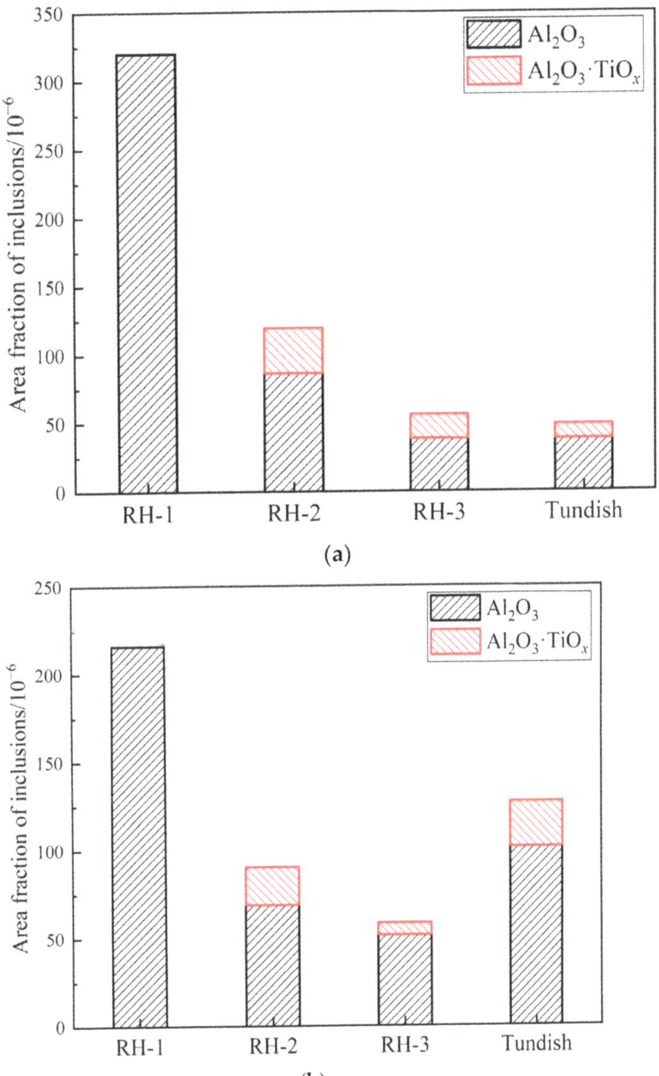

Figure 3. Change of the area fraction of inclusions in (**a**) Heat 1 and (**b**) Heat 2 during RH treatment and tundish processes.

IF steel in this plant is produced via the BOF–LF–RH–CC process, and the molten steel is in a state of peroxidation at the end of BOF. The oxygen in molten steel is continuously transferred to the ladle slag, resulting in the near-equilibrium conditions at the steel/slag interface. Moreover, no pre-deoxidation measures are taken before the RH refining. Consequently, the ladle slag oxidability is still at a high level despite taking the slag denaturalization treatment, and the T.Fe content in the ladle slag after RH is basically over 15%. Figure 4 shows the mean Ti percentage through the arithmetic average of Ti-containing inclusions. In this figure, the mean Ti percentage in Ti-containing inclusions increases gradually, especially more greatly from RH-end to tundish. It indicates that the reoxidation of molten steel occurs during the holding process. Simultaneously, the inclusion amount and the T.O content in molten steel for Heat 2 have a large increase in the tundish sample, indicating that the reoxidation is relatively serious under the actual producing conditions in this plant.

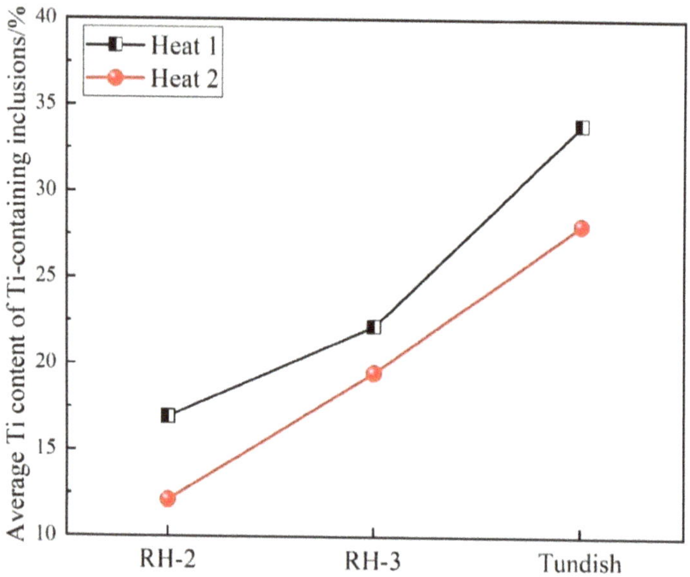

Figure 4. Transformation of inclusion composition during RH refining and tundish processes.

The unstable oxides in slag or refractory materials, including FeO, MnO, SiO_2 and other oxides; a source of oxygen in the added Ti-Fe alloy or sponge titanium; and the air inhalation into molten steel under the poor protection effect may all cause the reoxidation of molten steel, resulting in the increased content of the dissolved oxygen and total oxygen and a large loss of alloying elements in molten steel [14,19–21]. As can be seen from the change of N content in Table 1, the increase in N content in molten steel from RH-end to tundish is below 4 ppm. During the whole producing process, the N contents in molten steel do not obviously increase and are all below 20 ppm. Consequently, the reoxidation caused by the absorbed air is basically excluded.

In addition, the inclusion amount and the T.O content in molten steel for Heat 1 do not increase when the same sponge titanium is added into steel. Hence, under this experimental condition, the ladle slag with high oxidability is the main reason for the serious reoxidation of molten steel. During the RH refining process, the molten steel driven by argon gas has better stirring conditions. Even when the slag is highly oxidizing, inclusions can float up more quickly. Therefore, the ND and AF of inclusions in two heats gradually decrease during RH treatment. At the holding stage, compared with the alloying time in RH, the holding time is usually more than 2 times as long, and the stirring strength of molten steel becomes weak. Because of the worse conditions for the floating up of inclusions and the strong oxidized top slag, the molten steel may undergo serious reoxidation, leading to a significant increase in the ND and AF of inclusions in the tundish sample. For Heat 2, the T.Fe content in slag after RH reaches 18.1%, and the holding time is also more than 40 min. Before arriving at tundish, the oxygen in the ladle slag will transfer into the molten steel continuously, resulting in the deteriorated cleanliness of molten steel.

To sum up, in order to improve the cleanliness of molten steel, it is crucial to weaken the reoxidation by reducing the oxidability of slags and optimizing the steelmaking process during RH refining and tundish processes.

Figure 5 shows the size distribution of inclusions in Heat 1 and Heat 2 during RH treatment and tundish processes. From the figure, it can be seen that the ND of inclusions within 3 μm is the largest and the ND of inclusions above 5 μm has dropped to less than 2 mm^{-2}. Besides, the number of various inclusions appears to decrease with the increase in inclusion size. The number of Al_2O_3 and $Al_2O_3 \cdot TiO_x$ inclusions in Heat 1 keeps decreasing from the RH to tundish. However, in the tundish sample of Heat 2, the number of inclusions within 2 μm increases significantly owing to the serious reoxidation, which is also the main reason for the decrease in the mean size of Al_2O_3 inclusions. Meanwhile, it can be seen from the cumulative frequency distribution of inclusions that small-sized inclusions below 3 μm increase from around 77% before Ti addition to more than 81% after Ti addition. The long holding time between the removal of the ladle and the start of continuous casting gives inclusions the chance to grow and agglomerate. This results in a relatively strong increase in the percentage of the inclusions above 8 μm in tundish, which is a consequence of the maturation of the inclusion population. In other processes, the number of inclusions within 8 μm makes up almost 100%.

The Al_2O_3 cluster inclusions in IF steel are considered to be more harmful because of their larger size [22]. In the analyzing process of using ASPEX, many large-sized cluster-like inclusions may be defined as several inclusions, which will affect the accuracy of calculating the number of inclusions. In this study, a new method of determining the number of cluster inclusions is used to evaluate the cleanliness of IF steel. All Al_2O_3 and $Al_2O_3 \cdot TiO_x$ inclusions in the total detected area of a sample are assumed to be spherical, and then they are positioned on a map according to the location information of every inclusion. The relative size of the spherical dimension in the figure is determined depending on the size of inclusions, and multiple inclusions overlapping each other are considered as a cluster inclusion, and then the number of cluster inclusions can be counted.

The distribution of inclusions in the total sample detection area for Heat 1 and Heat 2 during RH treatment and tundish processes is shown in Figure 6. In the sample of RH-1, there are many small-sized inclusions aggregated together to form a cluster, including 38 clusters in Heat 1 and 21 clusters in Heat 2. During the period after Ti addition, the number of cluster inclusions decreases significantly because some larger inclusions have floated up and been removed. The single inclusion of forming the cluster is also larger in size, and even some inclusions are over 10 μm. After finishing the holding stage, compared with Heat 1, the number of cluster inclusions in the tundish sample of Heat 2 reaches 19, indicating that there are many cluster inclusions existing in the molten steel when the molten steel undergoes serious reoxidation. In the meantime, for Heat 1, the number of single inclusions in the tundish sample decreases obviously from Figure 6. However, the number of single inclusions below 4 μm in the tundish sample of Heat 2 has a significant increase, so the size of inclusions decreases significantly, as described in Figure 2. Additionally, the number of clusters obtained by this method is in good agreement with the ND and AF of inclusions in Figures 2 and 3, indicating that it is relatively reasonable.

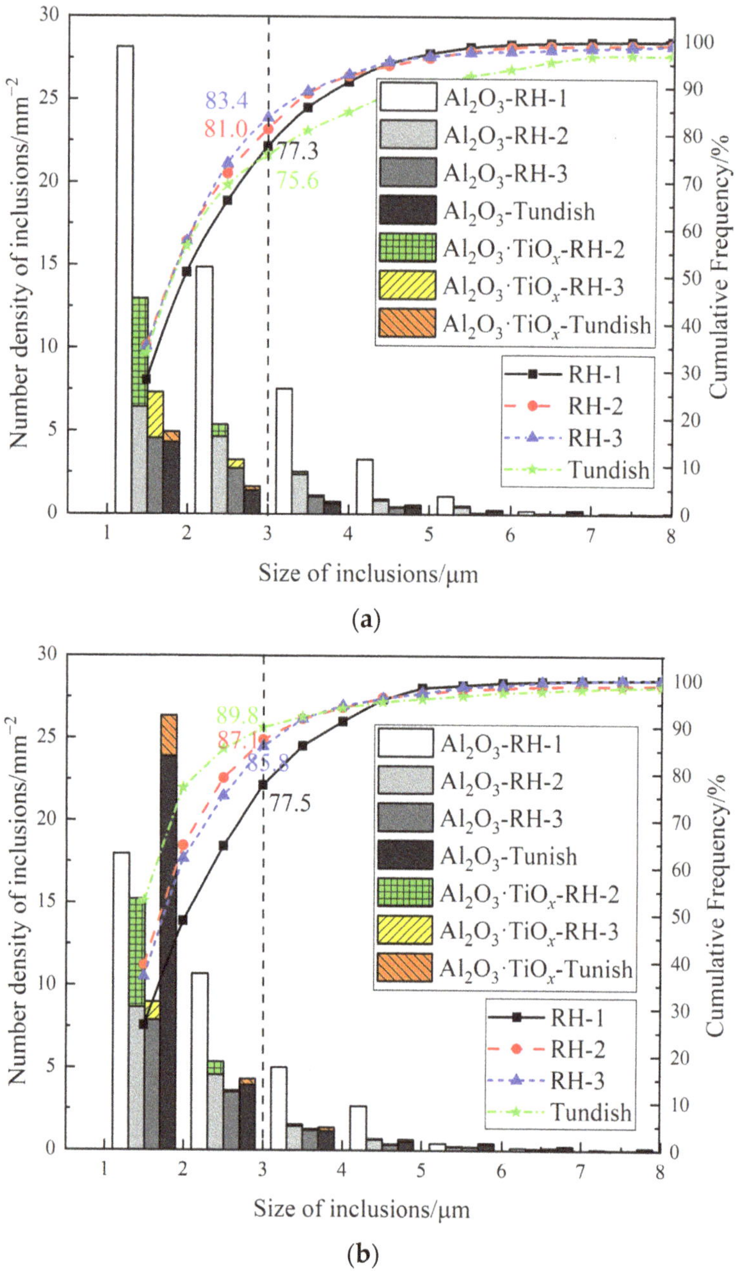

Figure 5. Size distribution of inclusions in (**a**) Heat 1 and (**b**) Heat 2 during RH treatment and tundish processes.

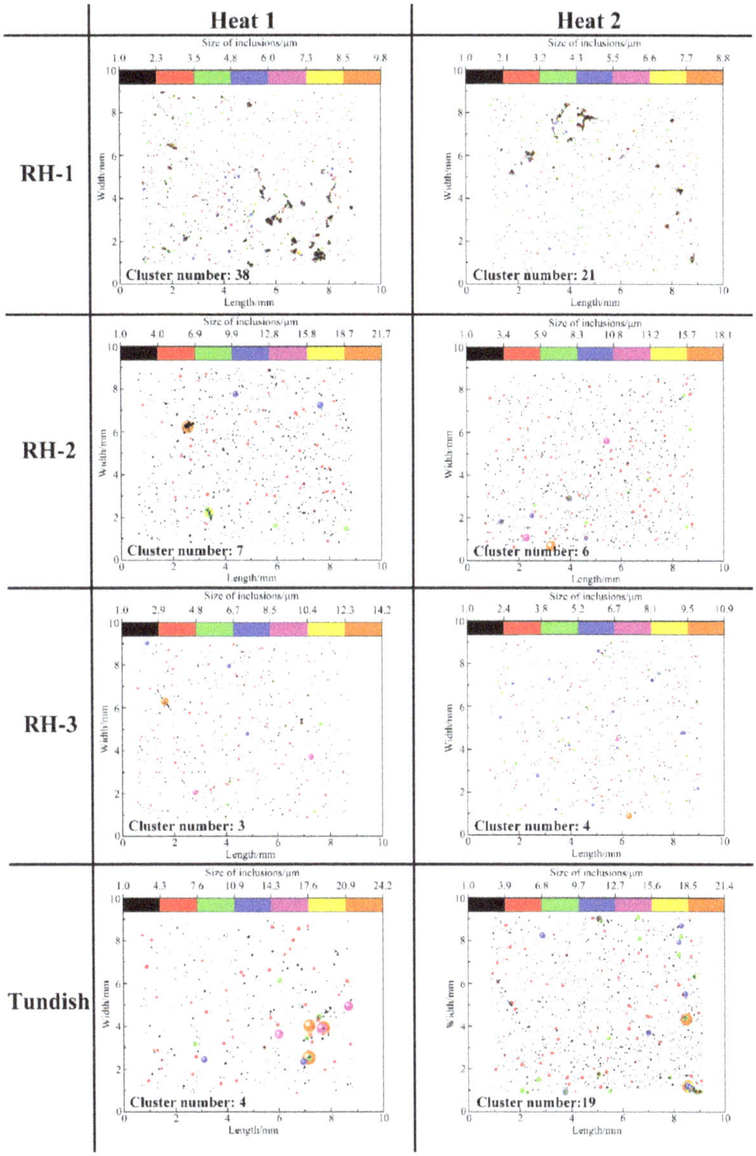

Figure 6. Distribution of inclusions in the total sample detection area for Heat 1 and Heat 2 during RH treatment and tundish processes.

3.2. Morphological Evolution of Inclusions

The morphological evolution of Al_2O_3 inclusions in molten steel from RH to tundish is depicted in Figure 7. Once the Al is added into the molten steel during the RH process, a large number of small-sized Al_2O_3 inclusions with different shapes are formed in molten steel, including spherical, bar-like, polygonal and other irregular shapes. These inclusions begin to aggregate together to form large Al_2O_3 clusters as shown in Figure 7 (RH-1). At around 3 min after Al addition, these small-sized Al_2O_3 inclusions cannot float up rapidly, and the ND of Al_2O_3 inclusions reaches more than 35 mm^{-2}. This indicates that it will take a longer time to further polymerize and grow to be removed for these inclusions.

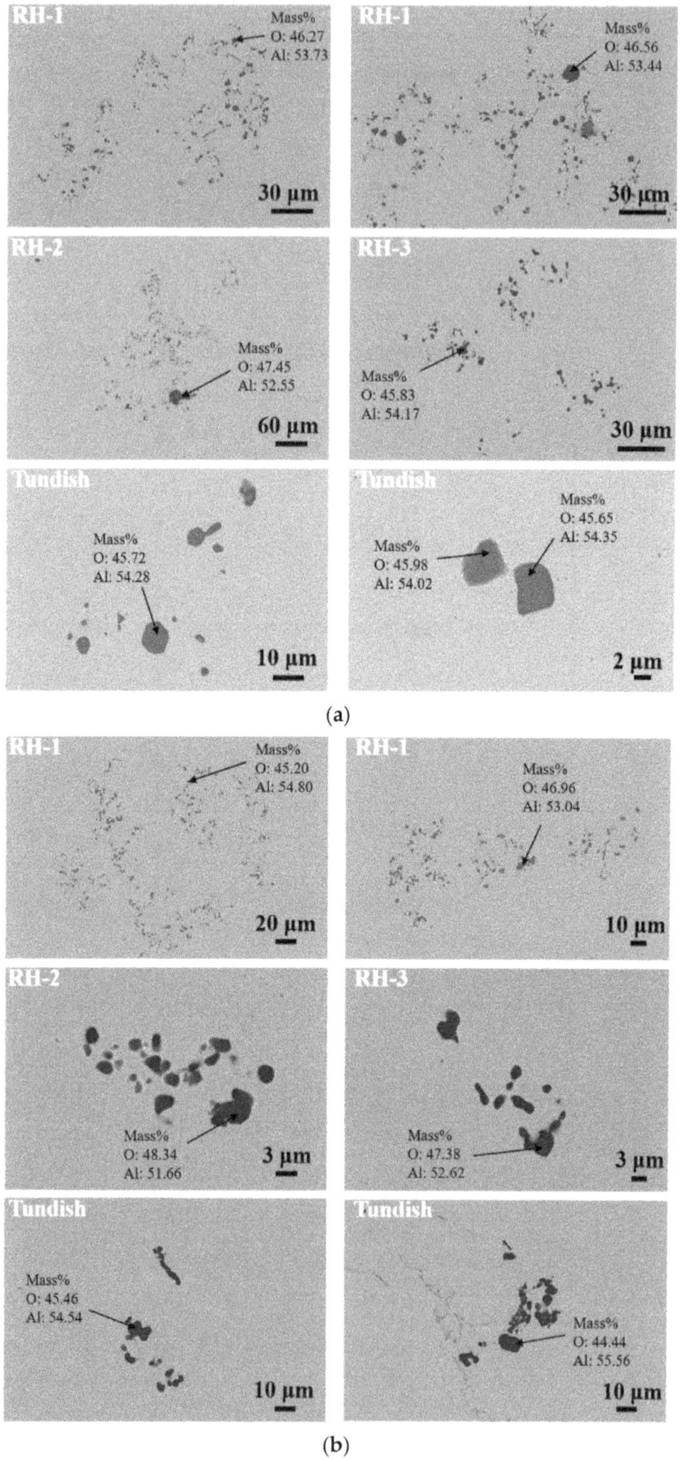

Figure 7. Morphologies of Al_2O_3 inclusions during RH treatment and tundish processes. (**a**) Heat 1, (**b**) Heat 2.

After Ti addition, there is still a certain amount of Al_2O_3 clusters in molten steel, as shown in Figure 7 (RH-2 and RH-3). Few Al_2O_3 inclusions can react with Ti in molten steel to form $Al_2O_3{\cdot}TiO_x$ inclusions, and most of the clusters are still pure Al_2O_3 inclusions. The size of the Al_2O_3 cluster decreases significantly at the end of RH, and the ND of Al_2O_3 inclusions in molten steel drops to less than 15 mm^{-2}, which is a combined result of the larger inclusions being removed more rapidly from molten steel through floatation and smaller inclusions precipitating. The decreasing trend of the T.O content in molten steel is consistent with the above evolution characteristic of inclusions, and it can be speculated that the Al deoxidation process at about 6 min after Ti addition could be completed because the oxygen level is close to the equilibration at the end of RH.

In the tundish sample, few cluster-like Al_2O_3 inclusions in molten steel of Heat 1 are found, and the Al_2O_3 inclusions mostly exist as single particles. Meanwhile, many Al_2O_3 clusters still exist in molten steel of Heat 2, as shown in Figure 7 (tundish). According to the analysis results in Section 3.1, this is mainly due to the serious reoxidation of molten steel caused by the ladle slag with high oxidation and the long holding time.

Figure 8 shows morphologies of $Al_2O_3{\cdot}TiO_x$ inclusions during RH treatment and tundish processes, indicating that there are mainly three types of $Al_2O_3{\cdot}TiO_x$ inclusions. The first type is similar to the shape of Al_2O_3 inclusions, including spherical and polygonal shapes, and the size is usually smaller than 5 μm, as shown in Figure 8a–d. The other two types of $Al_2O_3{\cdot}TiO_x$ inclusions are substantially spherical and relatively large in size, as shown in Figure 8e–l. Figure 9 shows the elemental mapping of $Al_2O_3{\cdot}TiO_x$ inclusions during RH treatment and tundish processes. In the first type, an Al-Ti-O corner is distributed on the surface of the Al_2O_3 inclusion, and the morphology and composition are nearly identical to pure alumina inclusions, as shown in Figure 9a,b. In the second type, inclusions with traces of Al and Ti are homogeneously distributed, as shown in Figure 9c–e, and some inclusions such as those in Figure 9c,e are surrounded by an Al_2O_3 layer with an Al-Ti-O core. In the third type, there are heterogeneous Al-Ti-O inclusions with an irregular Al_2O_3 core and an Al-Ti-O or TiO_x surrounding layer, as shown in Figure 9f,g, and the size is the largest of the three types.

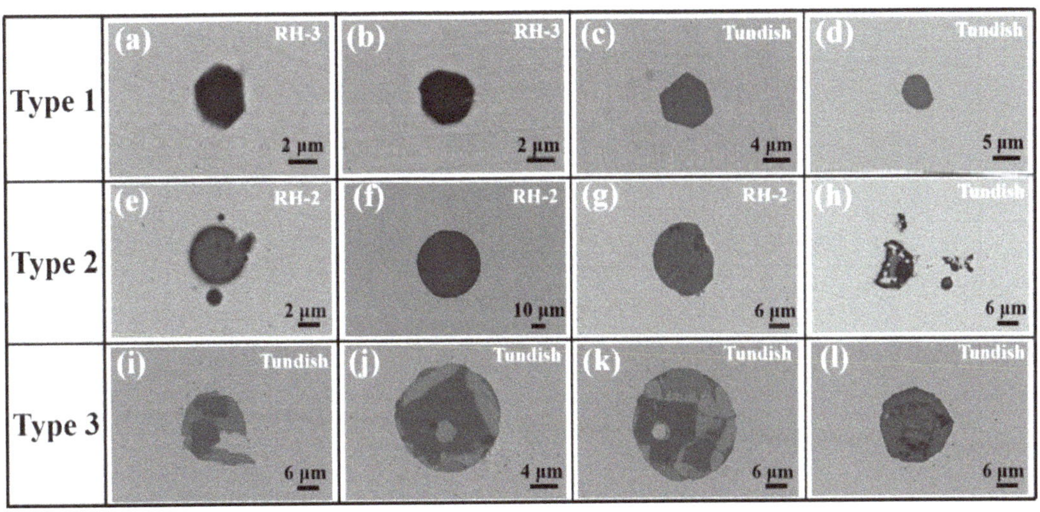

Figure 8. Morphologies of $Al_2O_3{\cdot}TiO_x$ inclusions during RH treatment and tundish processes. (**a–d**) Type 1 of $Al_2O_3{\cdot}TiO_x$ inclusions, (**e–h**) Type 2 of $Al_2O_3{\cdot}TiO_x$ inclusions, (**i–l**) Type 3 of $Al_2O_3{\cdot}TiO_x$ inclusions.

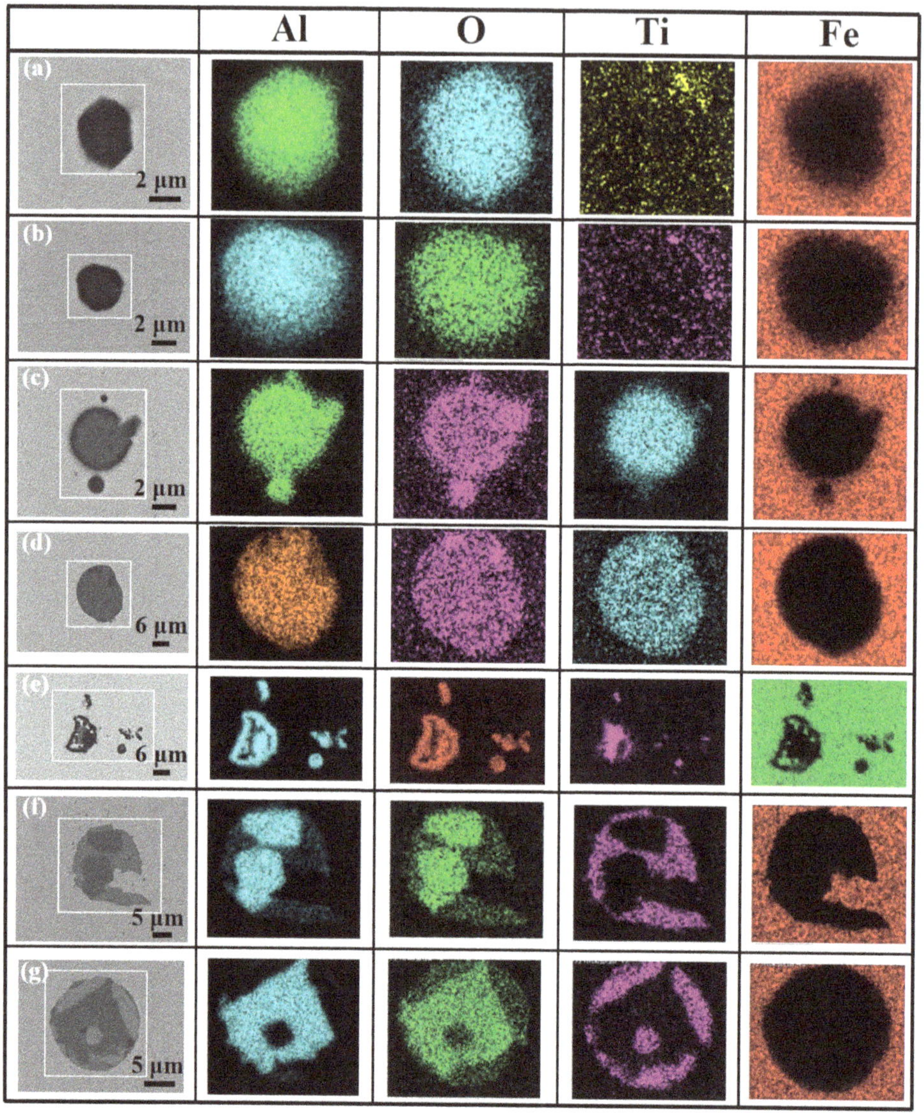

Figure 9. Elemental mapping of $Al_2O_3 \cdot TiO_x$ inclusions during RH treatment and tundish processes. (**a**,**b**) elemental mapping of Type 1 inclusions, (**c**–**e**) elemental mapping of Type 2 inclusions, (**f**,**g**) elemental mapping of Type 3 inclusions.

Similar morphologies of the first two types of $Al_2O_3 \cdot TiO_x$ inclusions were observed by Dorrer [19], Yang [23], Wang [24], Bai [25] and Yang [26]. The third type of $Al_2O_3 \cdot TiO_x$ inclusions, in which the Al-Ti-O is distributed in the outer part and the Al_2O_3 is distributed in the inner part, was also reported by Basu [14], Park [20], Wang [27], Doo [28], Qin [29] and Sun [30]. Furthermore, it is worth discussing the formation mechanism of different types of inclusions later.

The percentage of various $Al_2O_3 \cdot TiO_x$ inclusions during RH treatment and tundish processes is shown in Figure 10. From the figure, it can be seen that there are mainly two types of titanium-containing inclusions, namely Type 1 and Type 2, after Ti addition during the RH refining process. At 3 min after Ti addition, Type 1 of $Al_2O_3 \cdot TiO_x$ inclusions in two heats accounts for 83.8% and 93.2%, respectively. At the end of RH refining, the percentage of Type 1 inclusions decreases, and the percentage of Type 2 inclusions begins to increase. In the two tundish samples, Type 3 of $Al_2O_3{\cdot}TiO_x$ inclusions appears, but the number of Type 3 inclusions is significantly lower than that of the other two types. Meanwhile, the proportion of Type 1 inclusions further decreases, and the proportion of Type 2 inclusions further increases. This is mainly because under the good-stirring conditions in molten steel after Ti addition, the process of Ti reducing Al to form $Al_2O_3{\cdot}TiO_x$ inclusions has been proceeding, and thus more and more $Al_2O_3{\cdot}TiO_x$ inclusions with relatively uniform composition are generated. In addition, for the tundish samples, the total amount of inclusions in Heat 2 is larger, and the number of Type 1 and Type 2 inclusions in Heat 2 is significantly more than that in Heat 1, which is caused by the serious reoxidation process of molten steel. Therefore, the types of $Al_2O_3{\cdot}TiO_x$ inclusions generated in molten steel are not only related to the existence of Ti in molten steel, but also closely related to the reoxidation process of molten steel caused by the excessively high oxidability of top slag.

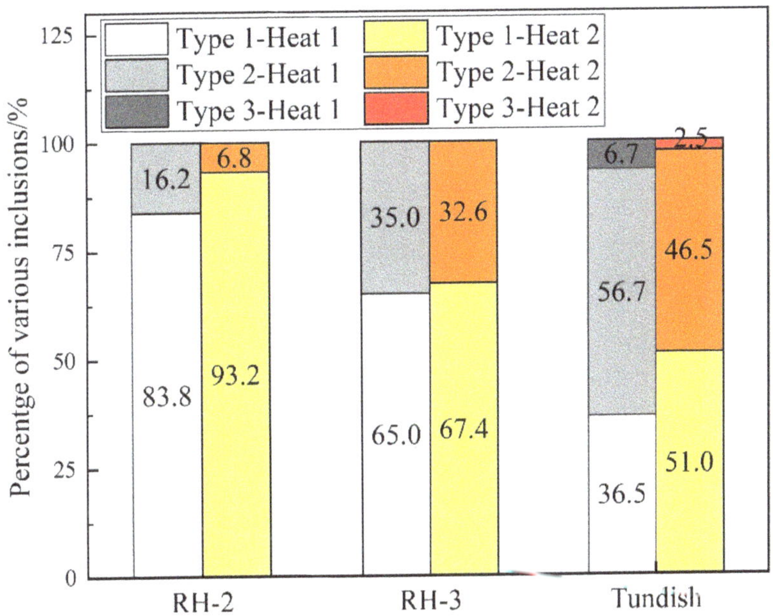

Figure 10. Percentage of various $Al_2O_3{\cdot}TiO_x$ inclusions during RH treatment and tundish processes.

The Ti/Al ratio in various $Al_2O_3{\cdot}TiO_x$ inclusions during RH treatment and tundish processes is presented in Figure 11. During the whole steelmaking process, the Ti content in Type 2 and Type 3 inclusions is significantly larger than that in Type 1 inclusions, and the Ti and Al contents in Type 1 inclusions almost keep constant. During RH treatment, the Ti and Al contents in Type 2 inclusions have a small fluctuation. After arriving at tundish, the Ti/Al ratio in Type 2 inclusions increases significantly owing to the increase in Ti content. In the tundish sample, the Ti/Al ratio of Type 3 inclusions among the three types of $Al_2O_3{\cdot}TiO_x$ inclusions reaches the maximum, indicating that the relative Ti content in Type 3 inclusions is at the largest level. Additionally, the Ti/Al ratio of Type 2 and Type 3 inclusions in the tundish sample of Heat 2 is larger than that in the tundish sample of Heat 1. This indicates that the Ti content in Type 2 and Type 3 inclusions will increase obviously when the molten steel encounters serious reoxidation.

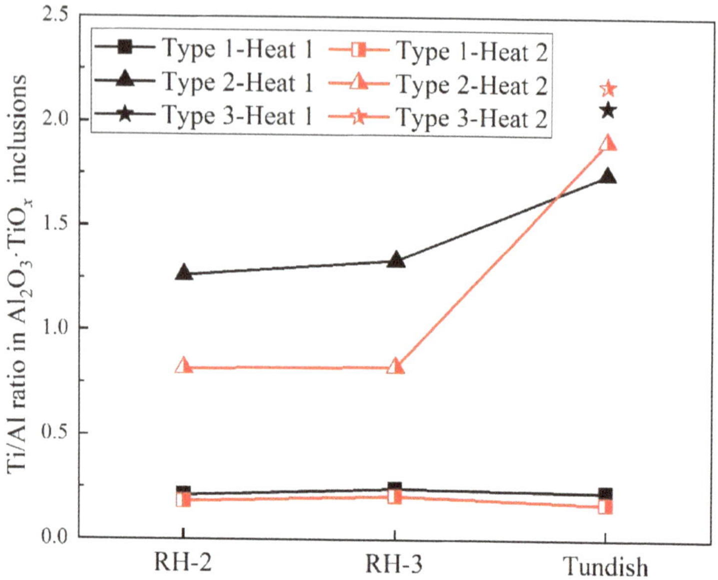

Figure 11. Ti/Al ratio in various $Al_2O_3 \cdot TiO_x$ inclusions during RH treatment and tundish processes.

3.3. Formation Evolution of Inclusions

Most research studies in this field have focused on the formation evolution of Al_2O_3 inclusions created in the laboratory through small furnace trials [31–33]. However, the formation of Al_2O_3 clusters in the industrial steelmaking process should be studied more completely. It is well known that nucleation, growth and removal of inclusions during the deoxidation process determine the size distribution of the inclusion population. The formation and removal mechanisms of Al_2O_3 inclusions during RH refining and tundish processes are depicted in Figure 12. Al_2O_3 inclusions below 1 μm can nucleate quickly after starting deoxidation. At around 3 min after Al addition, these tiny Al_2O_3 particles in molten steel grow, collide with each other and aggregate to form the large Al_2O_3 cluster as shown in Figure 7 (RH-1).

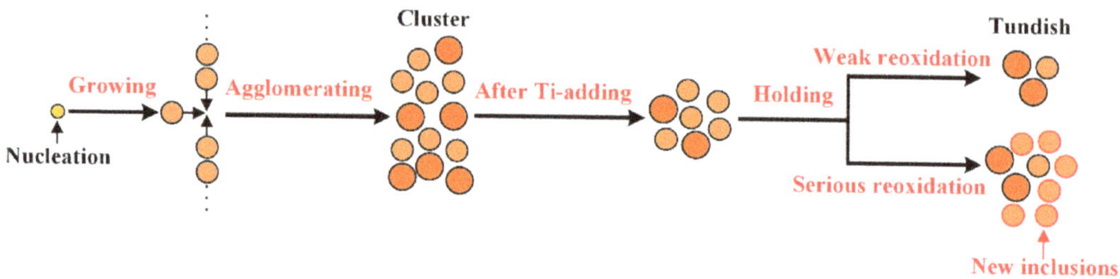

Figure 12. Formation and removal mechanisms of Al_2O_3 inclusions during RH refining and tundish processes.

During the pure circulation period after Ti addition, many large-sized Al_2O_3 inclusions have been removed, and some small-sized Al_2O_3 clusters still exist in molten steel, as shown in Figure 7 (RH-2 and RH-3). When the slag oxidability is relatively low with the weak reoxidation process during the holding process, the cleanliness of molten steel will not deteriorate, and most of the Al_2O_3 inclusions form single inclusion particles, as shown in the tundish sample of Figure 7a. Nevertheless, when the unstable oxide contents (FeO, MnO, SiO_2) in the ladle slag are large and the holding time is also relatively long, the molten steel will undergo serious reoxidation and the cleanliness of molten steel will deteriorate. There are many small-sized Al_2O_3 inclusions formed by the reoxidation process, so it can be seen in Figure 7b that there are still some Al_2O_3 clusters in the tundish sample.

The mechanism of the reoxidation process of molten steel by the top slag during the holding stage is shown in Figure 13. When the unstable oxide contents (FeO, MnO, SiO_2) in the top slag are at a high level, the oxygen can transfer continuously into molten steel at the steel/slag interface. Thus, the [Al] in molten steel combines with the oxygen at the steel/slag interface to form Al_2O_3 inclusions, according to Reaction (R1). In the meantime, the [Al] and [Ti] in molten steel may also react with the oxygen at the steel/slag interface, resulting in the formation of $Al_2O_3{\cdot}TiO_x$ inclusions, based on Reaction (R2). The morphologies of $Al_2O_3{\cdot}TiO_x$ inclusions formed by the reoxidation process are mainly the Type 1 and Type 2 inclusions as shown in Figure 8. The number of Type 3 $Al_2O_3{\cdot}TiO_x$ inclusions is relatively low as described in Figure 10, and the formation conditions are fairly limited during the reoxidation process, which will be explained in detail below. In addition, the formation of TiO_x inclusions is possible at the steel/slag interface based on Reaction (R3).

$$2[\mathrm{Al}] + 3[\mathrm{O}]^* = \mathrm{Al_2O_3} \tag{R1}$$

$$2[\mathrm{Al}] + [\mathrm{Ti}] + (x+3)[\mathrm{O}]^* = \mathrm{Al_2O_3} \cdot \mathrm{TiO}_x \tag{R2}$$

$$[\mathrm{Ti}] + x[\mathrm{O}]^* = \mathrm{TiO}_x \tag{R3}$$

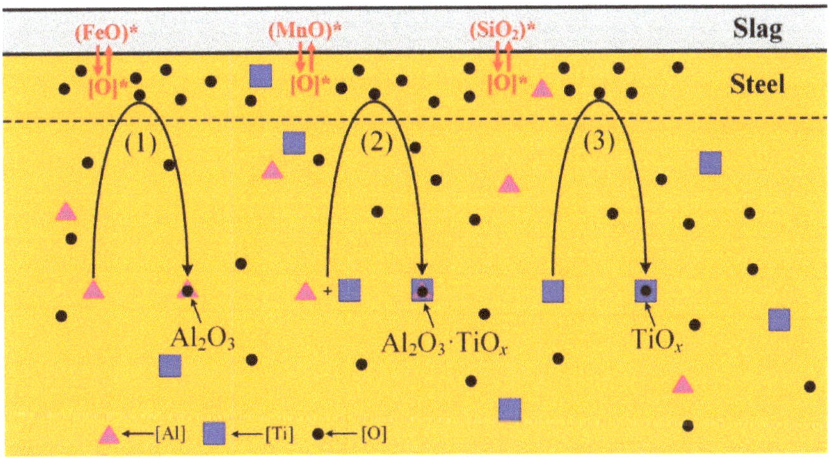

Figure 13. Mechanism of the reoxidation process of molten steel by the top slag during the holding stage.

Many scholars [23–30] have studied the formation mechanism of $Al_2O_3{\cdot}TiO_x$ inclusions, and their studies are good for analyzing the formation mechanisms of various $Al_2O_3{\cdot}TiO_x$ inclusions. The schematic of the process is depicted in Figure 14. The $Al_2O_3{\cdot}TiO_x$ inclusions formed by mechanisms I and II usually exist in the melt away from the steel/steel interface. After Ti feeding into molten steel, the formation of a local high [Ti] region provides the possibility of generation of Al-Ti-O or TiO_x inclusions based on Reactions (R4) and (R5). The reaction area between the high [Ti] region and the inclusion surface determines the final morphology of $Al_2O_3{\cdot}TiO_x$ inclusions. When only part of the area around Al_2O_3 inclusions is a high [Ti] region, the Al_2O_3 inclusion is only partially reduced by [Ti] with time, thus forming the Type 1 inclusions in Figure 8a–d, and the formation mechanism is depicted in Figure 14I.

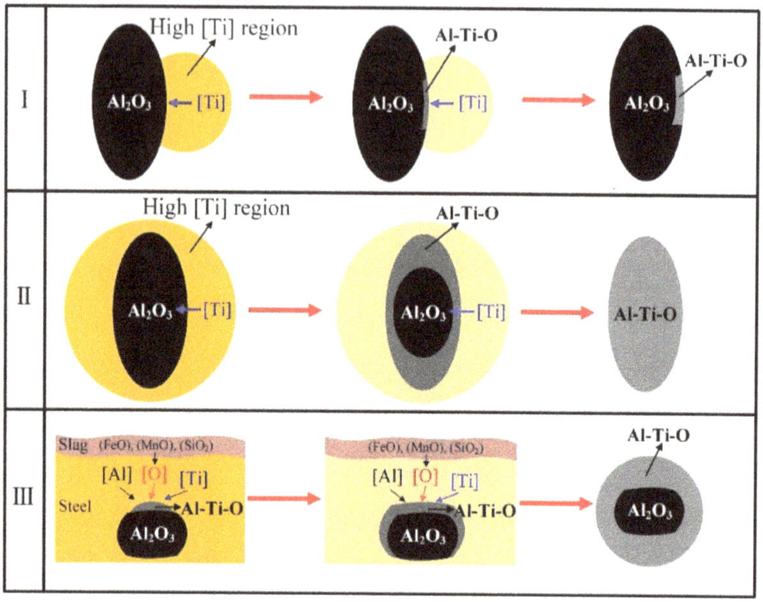

Figure 14. Formation mechanisms of $Al_2O_3{\cdot}TiO_x$ inclusions during RH refining and tundish processes. (**I**) first formation mechanism of $Al_2O_3{\cdot}TiO_x$ inclusions, (**II**) second formation mechanism of $Al_2O_3{\cdot}TiO_x$ inclusions, (**III**) third formation mechanism of $Al_2O_3{\cdot}TiO_x$ inclusions.

When the Al_2O_3 inclusion is completely covered by the high [Ti] region, the Al_2O_3 inclusion will gradually be reduced from the outside to the inside, thus forming the Type 2 inclusions as shown in Figure 8e–h. The formation mechanism of the inclusions is depicted in Figure 14II. After that, with the progress of reduction of Al_2O_3, a low [Ti] region around the inclusion exists. Thus, the $Al_2O_3{\cdot}TiO_x$ inclusion may transfer to Al_2O_3 from the surface to the inner part of the inclusion according to Reactions (R4) and (R5). As a result, the inclusion with an Al-Ti-O core surrounded by a Al_2O_3 layer forms finally, as shown in Figure 8e,h.

$$xAl_2O_3 + 3[Ti] = 3TiO_x + 2x[Al] \quad (R4)$$

$$5Al_2O_3 + 3[Ti] = 3Al_2TiO_5 + 4[Al] \quad (R5)$$

$$4[Al] + 3Al_2TiO_5 = 3[Ti] + 5Al_2O_3 \quad (R6)$$

$$2x[Al] + 3TiO_x = 3[Ti] + xAl_2O_3 \quad (R7)$$

The third type of $Al_2O_3{\cdot}TiO_x$ inclusions mainly exists in the tundish process. Previous studies [14,20,28,30] have pointed out that these inclusions are closely related to the reoxidation of molten steel, and the top slag with strong oxidability in this steel plant during the smelting process also provides conditions for the formation of such inclusions. The formation mechanism of the inclusions is depicted in Figure 14III. The ladle or tundish slag with high unstable oxide contents keeps transferring oxygen into molten steel, and the oxygen reacts with [Al] and [Ti] in molten steel to form the Al-Ti-O inclusion on the surface of the irregular Al_2O_3 inclusions according to Reaction (R2). Finally, the spherical inclusions partially or completely wrapped by the Al-Ti-O inclusion on the surface of pure Al_2O_3 inclusions appear in molten steel. Because of the continuous reoxidation process, newly formed Al-Ti-O inclusions accumulate constantly on the surface, so the size is also larger than the previous two types.

Interestingly, these irregular Al_2O_3 inclusions in Type 3 inclusions are usually larger in size as shown in Figure 8i–l, and they are unlikely to be generated from the reoxidation process. It is more likely that the Al_2O_3 inclusions accumulate and grow in the melt, then float up to the steel/slag interface and finally become the nucleation core of Type 3 inclusions. However, the number of Type 3 inclusions is relatively small, indicating that most of the mature Al_2O_3 inclusions have been absorbed by the top slag. According to the above analysis results, the number of Al_2O_3 inclusions formed by the reoxidation is significantly larger than that of $Al_2O_3 \cdot TiO_x$ inclusions formed by the reoxidation. Combining Figure 13 with Figure 14, the $Al_2O_3 \cdot TiO_x$ inclusions can not only form directly at the steel/slag interface, but also there is an indirect way of forming $Al_2O_3 \cdot TiO_x$ inclusions in the case of serious reoxidation of molten steel. Specifically, the Al_2O_3 inclusions formed by the reoxidation form at the steel/slag interface firstly, and then these new Al_2O_3 particles scatter in the melt. Eventually, these Al_2O_3 particles may be transferred into $Al_2O_3 \cdot TiO_x$ inclusions under the action of the high [Ti] region.

The equilibrium phase diagram for the Fe-Al-Ti-O system at 1873 K was calculated by using Factsage 7.2 with the FactPS, FToxid and FSstel databases in September 2020, as depicted in Figure 15. The black dot in Figure 15 represents the IF steel composition in the current work, indicating that Al_2O_3 is the only thermodynamically stable phase under the experimental conditions. From the figure, it can be seen that different types of Al-Ti-O inclusions have various modification mechanisms. The Type 1 and Type 2 of Al-Ti-O inclusions can be formed by the increase in the local [Ti] concentration, and the red solid arrow can represent the formation mechanisms of I and II in Figure 14. Meanwhile, the reoxidation of molten steel by the ladle or tundish slag leads to an increase in the dissolved oxygen and total oxygen in molten steel, and thus it is possible to generate directly or indirectly three types of Al-Ti-O inclusions, as presented by the blue solid arrow in Figure 15. In addition, with the uniformity of the molten steel composition, the Al-Ti-O inclusion may eventually be partially or fully turned into the Al_2O_3 inclusion by Al reducing, as shown by the green dashed arrow in Figure 15.

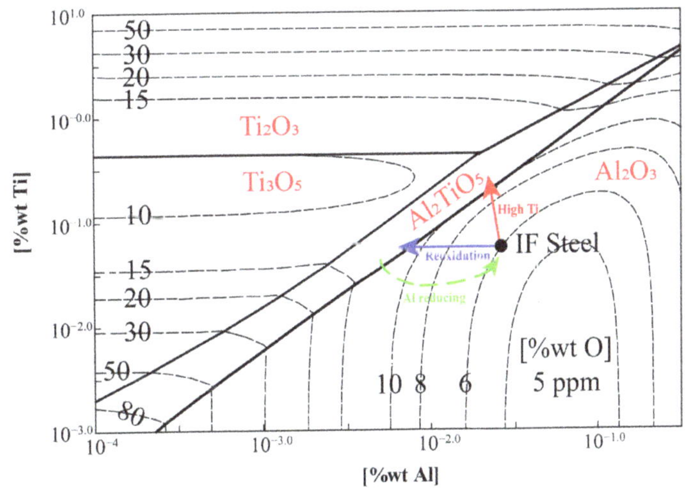

Figure 15. The equilibrium phase diagram for the Fe-Al-Ti-O system at 1873 K.

3.4. *Influence of Process Parameters on the Cleanliness of Molten Steel*

The secondary refining process is considered to be the main place for the removal of inclusions in molten steel, and the percentage of inclusions removed accounts for more than 70% [34]. Recently, to improve the cleanliness of molten steel, reducing the number of endogenous inclusions, improving the removal efficiency of inclusions and lessening the external inclusions caused by the reoxidation are dominant ways during the RH refining process. For this reason, studying the influence of relevant process parameters on the removal efficiency of inclusions is vital to improve the quality of the products.

The [O] content in molten steel before Al deoxidation is a key parameter that determines the amount of Al added and the number of inclusions in molten steel [15,34]. Figure 16 shows the relationship between the number of inclusions after RH and [O] content in molten steel before Al deoxidation. As shown in the figure, with the increase in the [O] content in molten steel, the number of inclusions in molten steel after RH tends to increase, and the cleanliness of molten steel becomes worse.

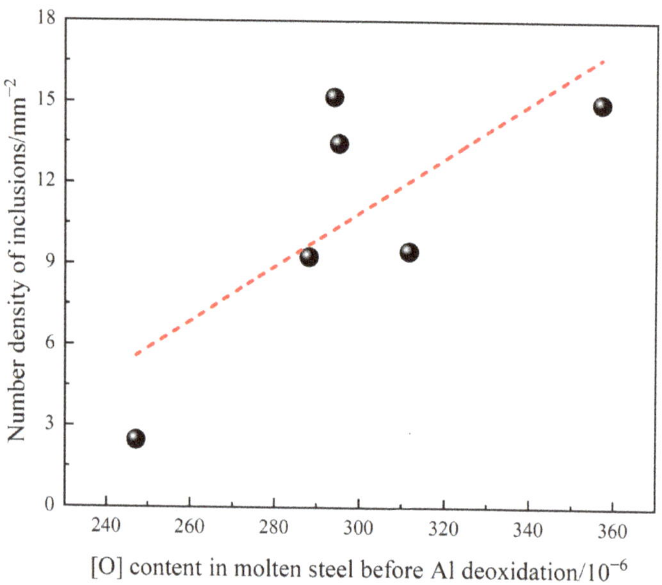

Figure 16. Relationship between the number of inclusions after RH and [O] content in molten steel before Al deoxidation.

According to the research of Wakoh [35], all dissolved oxygen was consumed for forming spherical inclusions, and the relationship between the oxygen content in molten steel and the number of inclusions in molten steel was obtained, as shown in Equation (3).

$$\frac{4}{3}\pi r^3 \rho_i N_i (\frac{3M_O}{M_i}) = (\frac{C_O}{10^6}) W_{Fe} \quad (3)$$

where r is the radius of inclusion, m; ρ_i is the density of inclusion, kg·m^{-3}; the density of Al_2O_3 inclusions is taken as 3.97×10^3 kg·m^{-3} [27]; M_O is the molar weight of oxygen, g·mol^{-1}; M_i is the molar weight of inclusion, g·mol^{-1}; C_O is the concentration of oxygen, 10^{-6}; and W_{Fe} is the total weight of molten steel, kg.

The relationship between the number of Al_2O_3 inclusions and [O] content in molten steel before Al deoxidation is depicted in Figure 17. From the figure, it can be seen that the number of Al_2O_3 particles formed in molten steel increases gradually with the gradual increase in dissolved oxygen in molten steel before Al deoxidation, and the increased amplitude is more obvious with the decreased radius of Al_2O_3 inclusions. In this paper, the average diameter of Al_2O_3 inclusions in these samples at 3 min after Al addition is only about 2 μm. Therefore, reducing the dissolved oxygen content before Al deoxidation has a significant effect on reducing the number of Al_2O_3 inclusions. Combining Figure 16 with Figure 17, it can be seen that reducing the dissolved oxygen content before Al deoxidation is beneficial to improve the cleanliness of molten steel.

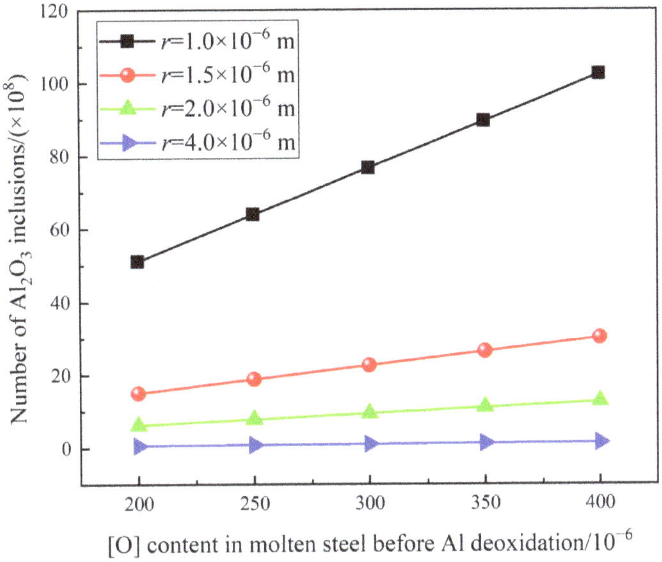

Figure 17. Relationship between the number of Al_2O_3 inclusions and [O] content in molten steel before Al deoxidation.

Figure 18 shows changes in the ND of inclusions, T.Fe content after RH and holding time in different heats. When the T.Fe content is above 15% and the holding time is more than 30 min, the ND of inclusions in molten steel is more than 9 mm^{-2}. However, when the holding time is less than 30 min or the T.Fe content of slag is below 15%, the number of inclusions in molten steel is relatively low. Therefore, it is beneficial to weaken the reoxidation process of molten steel and improve the cleanliness of molten steel by minimizing the holding time and reducing the oxidability of slag after RH.

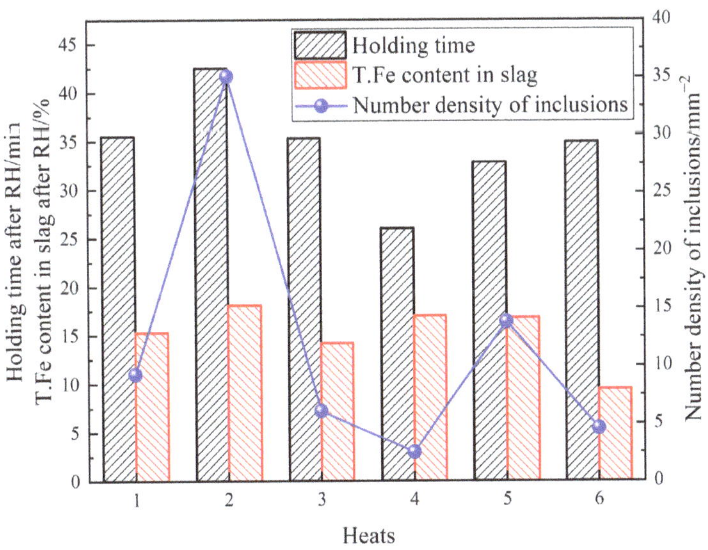

Figure 18. Changes in number density of inclusions, T.Fe content after RH and holding time in different heats.

4. Conclusions

Considering Ti-bearing IF steel produced via the BOF–LF–RH–CC process, the evolution of inclusions in molten steel during RH refining to tundish was analyzed through a systematic sampling experiment, and the influence of reoxidation process and process parameters on the cleanliness of molten steel was also studied. The results are summarized as follows:

(1) During RH treatment, the ND and AF of Al_2O_3 and $Al_2O_3 \cdot TiO_x$ inclusions decrease gradually. In the tundish samples, when the ND and AF of Al_2O_3 and $Al_2O_3 \cdot TiO_x$ inclusions especially from 1 to 2 μm increase significantly, the mean size of all inclusions shows a decreasing trend.
(2) The deteriorated cleanliness of molten steel is closely related to the serious reoxidation of molten steel caused by the slag with high oxidability during the holding process. Meanwhile, the number of clusters counted by the location maps shows basically the same trend with the ND and AF of inclusions.
(3) During the whole steelmaking process, Type 1 and Type 2 inclusions are the main two types of titanium-containing inclusions. In the tundish sample with serious reoxidation, there are still some small-sized Al_2O_3 clusters, and the number of Type 1 and Type 2 inclusions and Ti content of Type 2 and Type 3 inclusions will increase obviously.
(4) In the case of serious reoxidation, $Al_2O_3 \cdot TiO_x$ inclusions can not only form directly at the steel/slag interface, but also form indirectly: Al_2O_3 particles generated from reoxidation may be transferred into $Al_2O_3 \cdot TiO_x$ inclusions under the action of a local high [Ti] region.
(5) It is beneficial to weaken the reoxidation process and improve the cleanliness of molten steel by reducing the oxygen content in molten steel before Al deoxidation, minimizing the holding time and reducing the slag oxidability after RH.

Author Contributions: Conceptualization, B.Y. and J.L.; Methodology, B.Y. and J.L.; Software, B.Y.; Validation, B.Y., J.L. and J.Z.; Formal Analysis, B.Y.; Investigation, B.Y. and M.Z.; Resources, B.Y., J.L. and J.Z.; Data Curation, B.Y., J.H. and X.Y.; Writing—Original Draft Preparation, B.Y.; Writing—and Editing, B.Y.; Visualization, B.Y.; Supervision, J.L., J.Z. and M.Z.; Project Administration, J.L., J.Z., J.H. and X.Y.; Funding Acquisition, J.L., J.Z. and J.H. All authors have read and agreed to the published version of the manuscript.

Funding: The research was funded by the National Natural Science Foundation of China, grant number 51874028.

Data Availability Statement: Not applicable.

Acknowledgments: The authors would like to thank Pangang Group Xichang Steel and Vanadium Corporation Limited for providing technical and financial support. The authors gratefully acknowledge Pangang Group Research Institute Corporation Limited for the help and support with the ASPEX analysis.

Conflicts of Interest: The authors declare no conflict of interest. The funding sponsors had no role in the choice of the research project; the design of the study; the collection, analysis or interpretation of data; the writing of the manuscript; or the decision to publish the results.

References

1. Wang, C.; Nuhfer, N.T.; Sridhar, S. Transient behavior of inclusion chemistry, shape, and structure in Fe-Al-Ti-O melts: Effect of titanium source and laboratory deoxidation simulation. *Metall. Mater. Trans. B* **2009**, *40*, 1005–1021. [CrossRef]
2. Wang, C.; Nuhfer, N.T.; Sridhar, S. Transient behavior of inclusion chemistry, shape, and structure in Fe-Al-Ti-O melts: Effect of titanium/aluminum ration. *Metall. Mater. Trans. B* **2009**, *40*, 1022–1034. [CrossRef]
3. Deng, X.X.; Ji, C.X.; Zhu, G.S.; Liu, Q.M.; Huang, F.X.; Tian, Z.H.; Wang, X.H. Quantitative evaluations of surface cleanliness in IF Steel slabs at unsteady casting. *Metall. Mater. Trans. B* **2019**, *50*, 1974–1987. [CrossRef]
4. Matsuura, H.; Wang, C.; Wen, G.H.; Sridhar, S. The transient stages of inclusion evolution during Al and/or Ti additions to molten iron. *ISIJ Int.* **2007**, *47*, 1265–1274. [CrossRef]

5. Deng, X.X.; Liu, G.L.; Wang, Q.Q.; Liu, B.S.; Ji, C.X.; Li, H.B.; Shao, X.J.; Zhu, G.S. Effect of the weir structure in the tundish on the cleanliness of IF steels and elimination of spot-like defects in deep drawing automobile steel sheets. *Metall. Res. Technol.* **2020**, *117*, 609. [CrossRef]
6. Deng, X.X.; Ji, C.X.; Guan, S.K.; Wang, L.C.; Xu, J.F.; Tian, Z.H.; Cui, Y. Inclusion behaviour in aluminium-killed steel during continuous casting. *Ironmak. Steelmak.* **2019**, *46*, 1428420. [CrossRef]
7. Choucair, H.; Zhou, J.Z.; Ali Jawad, B.; Liu, L.P. Study of the fatigue failure of engine valve springs due to non-metallic inclusions. *SAE Int.* **2012**, *5*, 388–394. [CrossRef]
8. Murakami, Y. *Metal Fatigue: Effects of Small Defects and Nonmetallic Inclusions*, 2nd ed.; Elsevier: London, UK, 2002; pp. 681–682.
9. Story, S.R.; Goldsmith, G.E.; Klepzig, G.L. Study of cleanliness and castability in Ti-stabilized ultra low carbon steels using automated SEM inclusion analysis. *Metall. Res. Technol.* **2008**, *105*, 272–279. [CrossRef]
10. Salgado, U.D.; Weiß, C.; Michelic, S.K.; Bernhard, C. Fluid force-induced detachment criteria for nonmetallic inclusions adhered to a refractory/molten steel interface. *Metall. Mater. Trans. B* **2018**, *49*, 1632–1643. [CrossRef]
11. Lee, J.H.; Kang, M.H.; Kim, S.K.; Kim, J.H.; Kim, M.S.; Kang, Y.B. Influence of Al/Ti Ratio in Ti-ULC steel and refractory components of submerged entry nozzle on formation of clogging deposits. *ISIJ Int.* **2019**, *59*, 749–758. [CrossRef]
12. Bernhard, C.; Xia, G.; Karasangabo, A.; Egger, M.; Pissenberger, A. Investigating the influence of Ti and P on the clogging of ULC steels in the continuous casting process. In Proceedings of the 7th European Continuous Casting Conference, Duesseldorf, Germany, 27 June 2011.
13. Kimura, H. Advances in High-Purity IF Steel Manufacturing Technology. *Nippon. Steel Tech. Rep.* **1994**, *61*, 65–69.
14. Basu, S.; Chouahary, S.K.; Girase, N.U. Nozzle clogging behaviour of Ti-bearing Al-killed ultra low carbon steel. *ISIJ Int.* **2004**, *44*, 1653–1660. [CrossRef]
15. Lyons, C.; Kaushik, P. Inclusion characterization of titanium stabilized ultra low carbon steels: Impact of oxygen activity before deoxidation. *Steel Res. Int.* **2011**, *82*, 1394–1403. [CrossRef]
16. Gao, S.; Wang, M.; Guo, J.L.; Wang, H.; Zhi, J.G.; Bao, Y.P. Characterization transformation of inclusions using rare earth Ce treatment on Al-killed titanium alloyed interstitial free steel. *Steel Res. Int.* **2019**, *10*, 1900194. [CrossRef]
17. Gao, S.; Wang, M.; Guo, J.L.; Wang, H.; Zhi, J.G.; Bao, Y.P. Extraction, distribution, and precipitation mechanism of TiN–MnS complex inclusions in Al-killed titanium alloyed interstitial free steel. *Met. Mater. Int.* **2021**, *27*, 1306–1314. [CrossRef]
18. Yuan, B.H.; Liu, J.H.; Zhou, H.L.; Huang, J.H.; Zhang, S.; Shen, Z.P. Refining effect of IF steel produced by RH forced and natural decarburization process. *Chin. J. Eng.* **2021**, *43*, 1107–1115.
19. Dorrer, P.; Michelic, S.K.; Bernhard, C.; Penz, A.; Rössler, R. Study on the influence of FeTi-addition on the inclusion population in Ti-stabilized ULC steels and its consequences for SEN-clogging. *Steel Res. Int.* **2019**, *90*, 1800635. [CrossRef]
20. Park, D.C.; Jung, I.H.; Rhee, P.C.H.; Lee, H.G. Reoxidation of Al–Ti containing steels by $CaO–Al_2O_3–MgO–SiO_2$ slag. *ISIJ Int.* **2004**, *44*, 1669–1678. [CrossRef]
21. Jiang, M.F.; Zhang, Z.X.; Wang, D.Y.; Liu, J. Inclusions in characteristics of Ti/Al deoxidation steel and the analysis of nozzle clogging problem. *Ind. Heat.* **2011**, *40*, 60–63.
22. Deng, X.X.; Ji, C.X.; Cui, Y.; Tian, Z.H.; Yin, X.; Shao, X.J.; Yang, Y.D.; Mclean, A. Formation and evolution of macro inclusions in IF steels during continuous casting. *Ironmak. Steelmak.* **2017**, *44*, 739–749. [CrossRef]
23. Yang, W.; Li, S.S.; Li, Y.B.; Wang, X.H.; Zhang, L.F.; Liu, X.F.; Shan, Q.L. Evolution of inclusions in Ti-bearing ultra-low carbon steels during RH refining process. *Mater. Pro. Fund.* **2013**, 3–16. [CrossRef]
24. Wang, C.; Nuhfer, N.T.; Sridhar, S. Transient behavior of inclusion chemistry, shape, and structure in Fe-Al-Ti-O melts: Effect of gradual increase in Ti. *Metall. Mater. Trans. B* **2010**, *41*, 1084–1094. [CrossRef]
25. Bai, X.F.; Sun, Y.H.; Zhang, Y.M. Transient evolution of inclusions during Al and Ti additions in Fe-20 mass pct Cr alloy. *Metals* **2019**, *9*, 702. [CrossRef]
26. Yang, W.; Zhang, Y.; Zhang, L.F.; Duan, H.J.; Wang, L. Population evolution od oxide inclusions in Ti-stabilized ultra-low carbon steels after deoxidation. *J. Iron Steel Res.* **2015**, *22*, 1069–1077. [CrossRef]
27. Wang, M.; Bao, Y.P.; Cui, H.; Wu, H.J.; Wu, W.S. The composition and morphology evolution of oxide inclusions in Ti-bearing ultra low-carbon steel melt refined in the RH process. *ISIJ Int.* **2010**, *50*, 1606–1611. [CrossRef]
28. Doo, W.C.; Kim, D.Y.; Kang, S.C.; Yi, K.W. The morphology of Al-Ti-O complex oxide inclusions formed in an ultra low-carbon steel melt during the RH process. *Met. Mater. Int.* **2007**, *13*, 249–255. [CrossRef]
29. Qin, Y.M.; Wang, X.H.; Huang, F.X.; Ji, C.X. Behavior of non-metallic inclusions of IF steel during production process. *J. Northeast Univ. (Nat. Sci.)* **2015**, *36*, 1614–1618.
30. Sun, M.K.; Jung, I.-H.; Lee, H.-G. Morphology and Chemistry of Oxide inclusions after Al and Ti complex deoxidation. *Met. Mater. Int.* **2008**, *14*, 791–798. [CrossRef]
31. Marie-Aline, V.E.; Guo, M.X.; Zinngrebe, E.; Dekkers, R.; Proost, J.; Blanpain, B.; Wollants, P. Morphology and growth of alumina inclusions in Fe–Al alloys at low oxygen partial pressure. *Ironmak. Steelmak.* **2009**, *36*, 201–208. [CrossRef]
32. Jin, Y.; Liu, Z.Z.; Takata, R. Nucleation and growth of alumina inclusion in early stages of deoxidation: Numerical modeling. *ISIJ Int.* **2010**, *50*, 371–379. [CrossRef]
33. Van Ende, M.-A.; Guo, M.X.; Proost, J.; Blanpain, B.; Wollants, P. Formation and Morphology of Al_2O_3 inclusions at the onset of liquid Fe deoxidation by Al addition. *ISIJ Int.* **2011**, *51*, 27–34. [CrossRef]

34. Yang, G.W.; Wang, X.H.; Huang, F.X.; Wang, W.J.; Yin, Y.Q. Transient inclusion evolution during RH degassing. *Steel Res. Int.* **2014**, *85*, 26–34. [CrossRef]
35. Wakoh, M.; Sano, N. Behavior of alumina inclusions just after deoxidation. *ISIJ Int.* **2007**, *47*, 627–632. [CrossRef]

metals

MDPI

Article

Effect of Rare Earth Cerium Content on Manganese Sulfide in U75V Heavy Rail Steel

Chao Zhuo, Rui Liu, Zirong Zhao, Yulei Zhang, Xiaoshuai Hao, Huajie Wu * and Yanhui Sun

Collaborative Innovation Center of Steel Technology, University of Science and Technology Beijing, Beijing 100083, China; beikezhuochao@163.com (C.Z.); a_liurui@163.com (R.L.); zirong1216@163.com (Z.Z.); zhangyulei02111996@163.com (Y.Z.); 18401619871@163.com (X.H.); sun_yanhui@163.com (Y.S.)
* Correspondence: wuhuajie@ustb.edu.cn; Tel.: +86-138-1080-6121

Abstract: To study the effect of Ce on the morphology of manganese sulfide, we added different contents of Ce into U75V heavy rail steel. The composition and morphology of sulfide in steel were analyzed. The inclusions' number, size, and aspect ratio were analyzed by automatic scanning electron microscope ASPEX. The results show that the inclusions in heavy rail steel without Ce are elongated MnS and irregular Al-Si-Ca-O inclusions. With the increase of Ce from 52 ppm to 340 ppm, the composition of main inclusions changes along the route of Ce_2O_2S-MnS → Ce_2O_2S-MnS-Ce_2S_3 → Ce_2O_2S-Ce_3S_4-Ce_2S_3 → Ce_2O_2S-Ce_3S_4-CeS. Ce has a noticeable spheroidization effect on MnS, which can make inclusions finely dispersed. When Ce content is 139 ppm, the average size of inclusions is the smallest. The mechanism of Ce-modified MnS was discussed by combining experimental results with thermodynamic calculations. Finally, the effect of Ce treatment on inhibiting MnS deformation was verified by simulated rolling.

Keywords: U75V heavy rail steel; MnS; Cerium treatment; gleeble

1. Introduction

U75V high-speed heavy rail steel is fine pearlite steel with high strength, toughness, and wear resistance. It exhibits high sensitivity to white spot, however, it is difficult to completely remove hydrogen from this type of steel. According to previous studies [1,2], an increase in the S content of steel or the presence of MnS can reduce the diffusion coefficient of hydrogen in steel, thereby mitigating the damage caused by hydrogen. However, due to the excellent deformation ability of MnS, the elongation of MnS along the rolling direction during the rolling process has become one of the essential factors causing excessive inclusions and inconsistencies in ultrasonic flaw detection. The large-size and long-striped structure of MnS induce anisotropy in steel and significantly reduce the transverse properties of the material. MnS inclusions initiate crack formation, which occurs at the interface between sulfide and the matrix under stress. With an increase in the load, the crack expands gradually along with the striped MnS inclusions [3,4]. Therefore, to improve the mechanical properties of heavy rail steel, it is particularly vital to control the morphology, size, number density, and distribution of MnS [5].

At present, there are generally two methods to control the morphology of MnS during the steelmaking process. The first method involves the formation of a mass of oxide inclusions in the molten steel by adding titanium, magnesium, zirconium, and other elements to promote the heterogeneous nucleation of MnS precipitates. This method allows the formations of small and dispersed MnS inclusions. Oikawa et al. [6] found that the size of MnS inclusions decreased significantly after adding Ti to Fe-0.1C-1Mn-0.02S steel. They proposed that the liquid nucleus of (Ti, Mn) O formed at the solid-liquid interface of steel serves as the heterogeneous nucleation site of MnS droplets. Xie et al. [7] reported that when the Mg content of 16MnCrS5 steel was 35–42 ppm, numerous fine spindle-shaped or spherical composite inclusions with MgO-Al_2O_3 as the core and MnS as the shell were

Citation: Zhuo, C.; Liu, R.; Zhao, Z.; Zhang, Y.; Hao, X.; Wu, H.; Sun, Y. Effect of Rare Earth Cerium Content on Manganese Sulfide in U75V Heavy Rail Steel. *Metals* **2022**, *12*, 1012. https://doi.org/10.3390/met12061012

Academic Editor: Roumen Petrov

Received: 16 May 2022
Accepted: 11 June 2022
Published: 14 June 2022

formed in molten steel. The deformation degree of the inclusions after simulated rolling and forging was also small. Lu et al. [8] studied the distribution and morphology of MnS inclusions in Zr-containing non-quenched and tempered steel ingots and forging rods. Their results showed that when Zr content is 66 ppm, MnS is spherical or angular under two-dimensional and three-dimensional observation, and the distribution is uniform. The second method is the addition of calcium or rare-earth elements to S-containing steel to form smaller and lesser deformed sulfides in order to control the aspect ratio (the ratio of length to width of an inclusion) of inclusions. Another study [9] showed that the addition of a calcium alloy to steel can promote the transformation of sulfide from the striped to spindle structure, and that the ratio of calcium to sulfur content in steel (w(Ca)/w(S)) has a significant influence on the shape and length of the inclusions. When w(Ca)/w(S) > 0.2, sulfide transforms to form a spindle structure, the average size of the inclusions decreases with the increase in w(Ca)/w(S). Although MnS morphology control has been widely investigated [10–12], strategies to MnS formation are still limited. For example, Ti and Zr are not suitable for steel with a high oxygen content. Toward steel with low basicity refining slag, added Ca replaces Al_2O_3 in the refining slag and increases the Al content of steel to generate Al_2O_3 inclusions, which may lead to nozzle clogging [13]. The atomic size of rare earth elements is large and they easily lose their outer electrons, resulting in high activity. Thus, rare earth elements serve as good purifying agents and inclusion modifiers.

Cerium (Ce, a rare earth element), exhibits high activity and readily reacts with oxygen and sulfur in molten steel to form oxides and sulfides with a high melting point. Moreover, Ce can refine the grain size and improve the mechanical properties of the steel, and has a certain effect on harmful impurity elements such as phosphorus and arsenic [14–18]. Ren et al. [19] performed a thermodynamic calculation and found that the evolution order of inclusions was $Al_2O_3 \rightarrow CeAlO_3 \rightarrow Ce_2O_2S \rightarrow Ce_2O_2S + CeS$ when the Ce content of ultra-low carbon Al deoxidized steel increased from 0% to 0.028%, which was verified by laboratory experiments. Liu et al. [20] found that the primary inclusions changed from $MgO{\cdot}Al_2O_3$-MnS to $CeAlO_3$-MgO-MnS, Ce_2O_2S-MgO-MnS, and Ce_2O_2S-MnS after the addition of 0.014%, 0.024% and 0.037% Ce to EH36 steel, respectively. The addition of 0.024% Ce can inhibit the precipitation of pure MnS because Ce has a strong desulfurization ability. Liu et al. [21] found that 0.015% of added Ce could transform irregular MnS and Al_2O_3 inclusions in medium-carbon low-alloy steel to spherical Ce_2O_2S and decrease the inclusion size. Liu et al. [22] applied spring steel as fasteners of high-speed rails; they found that the total oxygen content (T.O.) was at the lowest level when the Ce content was 0.045–0.065 wt%. Oversized and irregular sulfides and oxide inclusions transformed into rare earth oxides, sulfides, or oxygen sulfides that were spherical and had a size of3 μm. At present, the effect of Ce content on the inclusions in non-aluminum deoxidized U75V heavy rail steel has not been reported, and the mechanism underlying the modification effect of Ce on MnS in heavy rail steel remains to be studied.

2. Materials and Methods

The experiment was completed in a tubular Si-Mo resistance furnace. A 300 g U75V steel sample, taken from the casting billet and cut into small pieces, was put into a MgO crucible with an inner diameter of 40 mm and an outer diameter of 48 mm. A graphite crucible was coated on the MgO crucible to protect the furnace. The schematic diagram of the resistance furnace and the composition of molten steel are shown in Figure 1 and Table 1.

Table 1. Chemical composition of U75V steel (%).

Composition	**C**	**Si**	**Mn**	**P**	**S**	**V**	**Als**
Content	0.77	0.64	0.89	0.02	0.007	0.04	0.0015

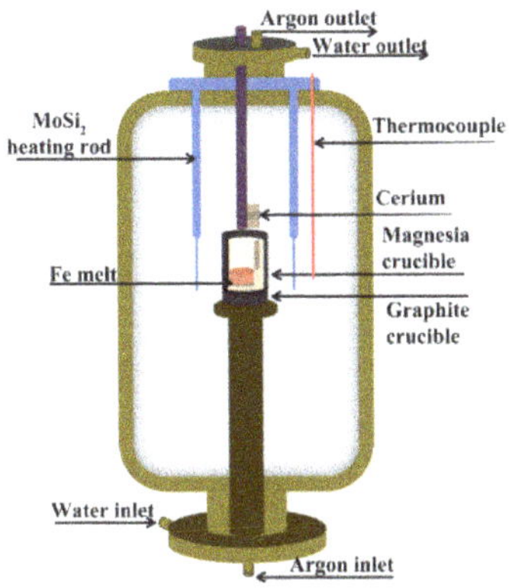

Figure 1. Schematic diagram of silicon molybdenum tube furnace.

The schematic diagram of the experimental process is shown in Figure 2a. After the crucible was placed, the upper lower ends of the furnace were closed; the temperature of the resistance furnace was set as shown in Figure 2a. When the temperature reached 1600 °C, it was held for 30 min to ensure that the steel blocks were melted and completely homogenized. Then, Ce was wrapped with pure iron foil and added to the molten steel without stirring. Five levels of Ce content, 52 ppm, 139 ppm, 171 ppm, 256 ppm, and 340 ppm, were used to study the effect of Ce content on inclusions in heavy rail steel. After holding for 15 min, the temperature began to decrease. When the temperature dropped to 700 °C, the power was cut off, and then the resistance furnace was cooled to room temperature by furnace cooling. In total, 2.5 L/min argon was injected throughout the experiment.

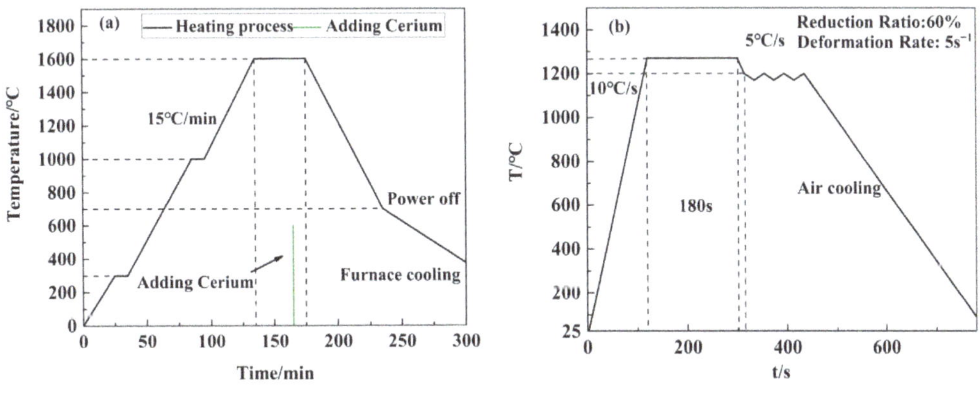

Figure 2. Temperature setting, (**a**): resistance furnace; (**b**): gleeble−3500 simulated rolling mill.

The ingot was cut into a cylindrical sample of ∅8 × 15 mm, which was used for the rolling simulated experiment. The temperature system refers to the production process of a plant, as shown in Figure 2b. The deformed samples were cut, polished, and observed with ASPEX. The simulation rolling experiment was completed on gleeble-3500 thermal simulation testing machine (Dynamic Systems Inc., Poestenkill, NY, USA).

The Ce content was measured by the ICP method. The composition, quantity, size, and aspect ratio of inclusions were analyzed by an ASPEX automatic scanning electron microscope (FEI, Hillsboro, OR, USA) after the ingot was ground and polished. The sampling position, analysis surface, and random observation position are shown in Figure 3, which reduces the influence of S element segregation inside the ingot. The scanning area of each sample was about 10 mm^2, and the minimum size of inclusions scanned was 1 μm. The inclusion morphology was observed by a ZEISS electron microscope equipped with an EDS (Energy Dispersive Spectrometer) model of Gemini SEM 500 (Zeiss, Niedersachsen, Germany). The extraction of inclusions was conducted by electrolysis with anhydrous organic solution, and the electrolyte was 10% AA solution (a mixed solution of 1% tetramethylammonium chloride-10% acetylacetone-89% methanol). Figure 4 is the schematic diagram of the electrolytic cell. The current density was 0.04 A/cm^2 and the temperature was 0~5 °C. The electrolysis time was 4 h. After electrolysis, the filtered membrane was filtered with Polytetrafluoroethylene (PTFE) filter membrane. After spraying gold on the filtered membrane, the three-dimensional characteristics of inclusions were observed by field emission scanning electron microscopy. Table 2 shows the Ce content in each sample, where C0 is the control group without Ce addition.

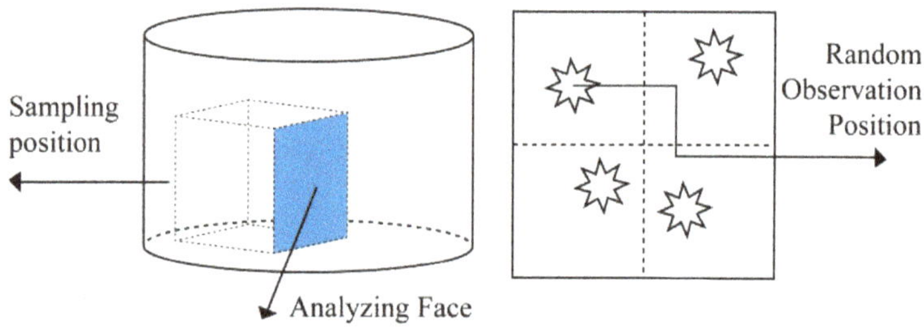

Figure 3. Sampling place, analyzing face, and random observation position.

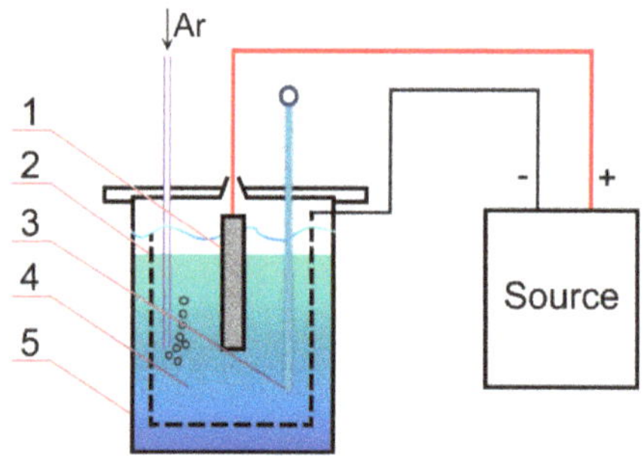

Figure 4. Schematic diagram of inclusion extraction (1—anode; 2—cathode; 3—thermometer; 4—electrolytes; 5—beaker).

Table 2. Ce content of each sample (mass%).

No.	C0	C1	C2	C3	C4	C5
Ce	0	0.0052	0.0139	0.0171	0.0256	0.0340

3. Results and Discussion

3.1. Composition and Morphology of Inclusions

Figure 5 shows the morphology and elemental mapping profiles of typical inclusions with different Ce contents. Figure 5a,b shows the control group C0 without Ce treatment. The inclusions are mainly large-scale irregular striped MnS and spherical Al_2O_3-SiO_2-CaO inclusions. After Ce treatment, the inclusions in steel convert to oxygen sulfides or Ce sulfides, which is in agreement with the result reported by Adabavazeh et al. [15] and Gao et al. [23]. Because of the low wettability and large contact angle between Ce-containing inclusions and molten steel [15], the inclusions in molten steel are ellipsoidal, and their size is markedly reduced. When the Ce content was 52 ppm, the primary inclusion type was Ce_2O_2S that formed the core, with a small amount of MnS composite inclusions precipitated on the surface, as shown in Figure 5c. In addition to the Ce_2O_2S-MnS composite inclusions, Ce_xS inclusions were observed when Ce content was 139 ppm and 171 ppm, as shown in Figure 5d. However, when the Ce content was approximately 256 and 340 ppm, the typical inclusions were single Ce_2O_2S inclusions and composite Ce_2O_2S and Ce_xS inclusions, as shown in Figure 5e,f. MnS was rarely precipitated on the surface of Ce_2O_2S, because with the increase in the Ce content, a large amount of S is consumed by the Ce_2O_2S and Ce_xS inclusions, causing the actual concentration of [S][Mn] in molten steel to not reach the equilibrium concentration.

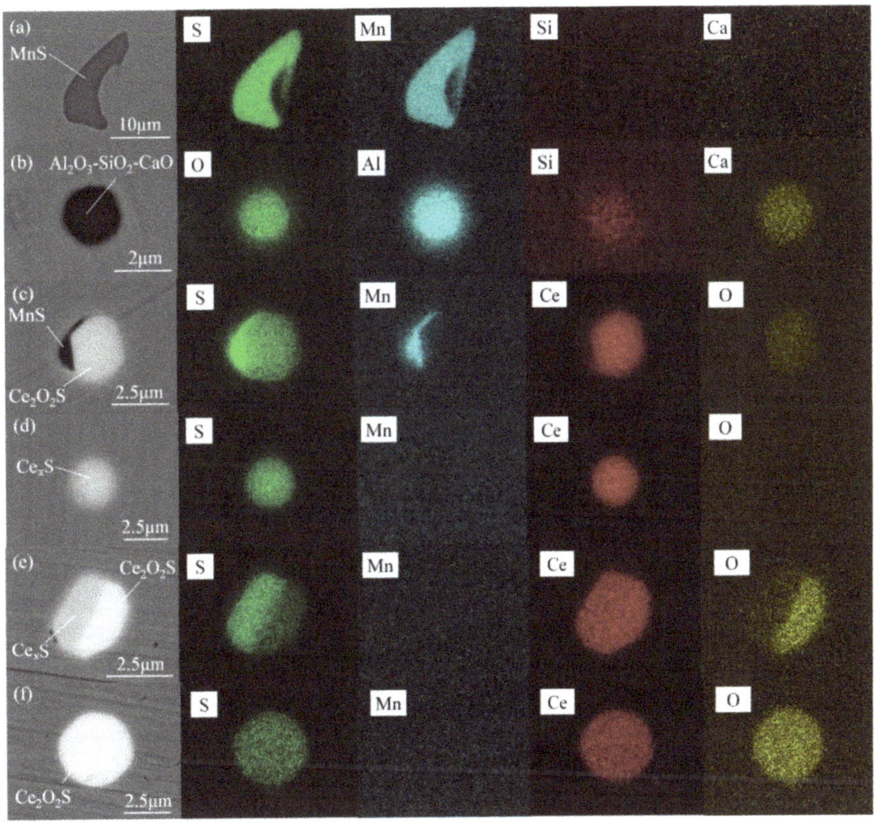

Figure 5. Morphology and elemental mapping of typical composite inclusions: (**a**,**b**) 0 ppm, (**c**) 52 ppm, (**d**) 139 ppm, 171 ppm, (**e**,**f**) 256 ppm, 340 ppm.

The three-dimensional morphology of typical inclusions with different Ce contents is shown in Figure 6. Figure 6a shows the morphology of MnS in the steel without Ce addition; it can be seen that MnS is irregular and large. When the Ce was 52 ppm, the typical inclusion was an ellipsoidal composite, with Ce_2O_2S as the core, and a small amount of MnS precipitates on the surface. No single irregular MnS precipitates were observed, as shown in Figure 6b. Figure 6c,d shows that the Ce_2O_2S-MnS-Ce_xS composite inclusions and the Ce_2O_2S-MnS inclusions appear when the Ce content of molten steel was 139 and 171 ppm, in addition to near-spherical Ce_xS inclusions. When the Ce content was 256, and 340 ppm, the inclusions in the steel were mainly single Ce_2O_2S and Ce_2O_2S-Ce_xS, as shown in Figure 6e,f. It is worth mentioning that in Figure 6c, the inclusion comprised Ce_2O_2S as the core, with the surface precipitation of MnS and Ce_xS composite inclusions. Theses type of inclusions are not observed in the two-dimensional morphology, indicating that three-dimensional morphology can more accurately reveal the type of inclusions. It can be found that the variation trend of inclusions is similar to that illustrated in Figure 5.

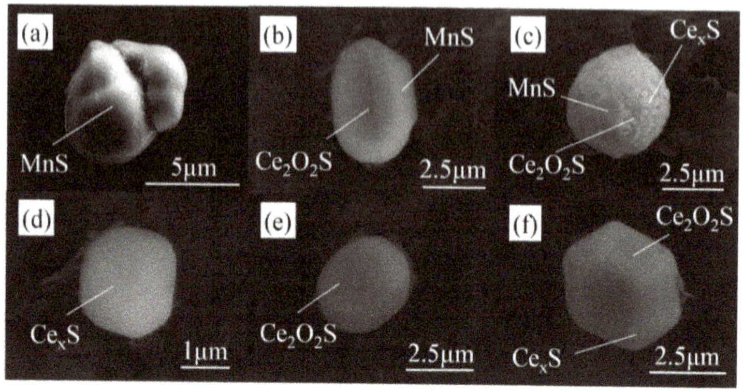

Figure 6. Three-dimensional morphology of typical inclusions in steel samples: (**a**) 0 ppm, (**b**) 52 ppm, (**c**,**d**) 139 ppm, 171 ppm, 256 ppm, (**e**,**f**) 340 ppm.

The typical inclusion types of each sample obtained after electron microscopy were classified and counted; the results are summarized in Table 3. When the Ce content was 52 ppm, the striped MnS disappeared. Nonetheless, the Ce content must be optimized on the basis of the size, number density, and aspect ratio of inclusions.

Table 3. Types of inclusions in different heats.

Typical Inclusions	Si-Al-Ca-O	MnS	Ce_2O_2S-MnS	Ce_xS	Ce_2O_2S	Ce_2O_2S-Ce_xS	Ce_2O_2S-Ce_xS-MnS
C0	√√	√√	–	–	–	–	–
C1	–	–	√√	–	–	–	–
C2	–	–	√√	√	–	–	√
C3	–	–	√√	√	√	√	√
C4	–	–	√	√	√	√√	–
C5	–	–	√	√	√	√√	–

√√: the main type of inclusions, √: a small number of inclusions.

3.2. *Number Density, Size, and Aspect Ratio of Inclusions*

To quantitatively characterize the inclusions in the steel sample, the number density, average size, and aspect ratio of the inclusions were calculated, the results are shown in Figures 7 and 8.

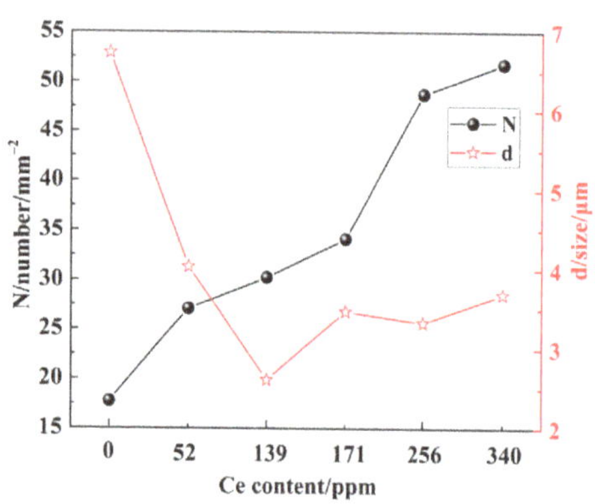

Figure 7. Number density and the average size of inclusion.

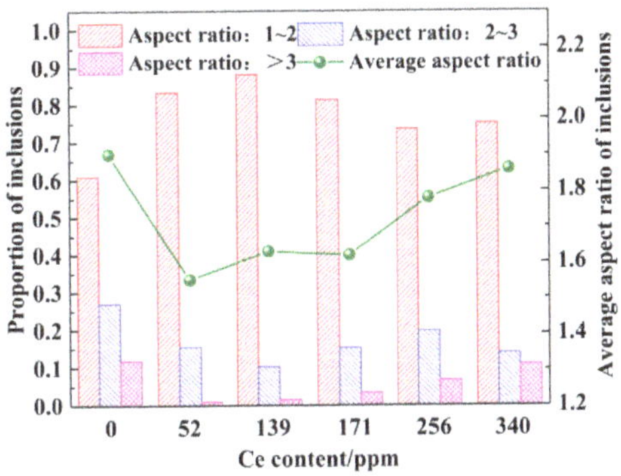

Figure 8. Aspect ratio of inclusions.

Figure 7 shows the changes in the number density and size of inclusions in different samples. Ce addition was found to significantly increase the number density of the inclusions in molten steel. When Ce was not added, the number density of the inclusions was 17.74/mm^2. With an increase in the Ce content, the number density of the inclusions increased, reaching a maximum of 51.76/mm^2 at 340 ppm Ce. In addition, Ce addition can significantly reduce the size of the inclusions. The average size of the inclusions before Ce addition was 6.74 μm, whereas after Ce addition it was 3.22–4.16 μm. Unlike the number density, the average size of the inclusions first decreased and then increased with an increase in the Ce content, which agrees with the result of Luo [24] and Wang [25]. Because of the change of free energy, the free energy of O and S binding in Ce and steel is considerably lower than that of S and Mn binding; thus, rare earth oxygen sulfides are easily generated [26]. Moreover, the melting point of rare earth sulfides is higher than that of MnS, and they precipitate before MnS during the solidification process. The continuous consumption of S reduces the activity of S in steel, which significantly affects the combination of Mn and S, thereby decreasing the core of heterogeneous nucleation, and reducing the probability of the transformation of single-particle MnS inclusions to large MnS inclusions under high supersaturation conditions. When the Ce content was high or excessive, the addition of excessive Ce enhances the binding ability of rare earth elements to the formation elements of the inclusions, resulting in the formation of a large number of rare earth inclusions. The inclusion collision and aggregation probabilities increase sharply; therefore, the inclusion size increases gradually.

The effect of the Ce content on the aspect ratio of the inclusions is shown in Figure 8. The average aspect ratio of the inclusions decreased significantly after Ce addition. The aspect ratio of the inclusions without Ce was 1.9, whereas for the inclusions with Ce it was between 1.55 and 1.86. The average aspect ratio of the inclusions increases with the Ce content. The proportion of the inclusions with an aspect ratio between one and two increased significantly after Ce treatment. The proportion of inclusions with an aspect ratio between two and three and >three decreased, indicating that Ce addition induces an apparent spheroidization effect on the inclusions. However, when the Ce content is excessive, i.e., more than 256 ppm, the proportion of the inclusions with an aspect ratio between one and two decreased gradually, whereas the proportion of the inclusions with an aspect ratio between two and three increases gradually. This trend is observed because inclusion collision and aggregation lead to an irregular shape and size enlargement. It indicates that excessive Ce content is not conducive to the dispersion and fine control of the inclusions. According to the number density, size, and aspect ratio of the inclusions, many fine and dispersed ellipsoidal inclusions can be generated at 139 ppm Ce.

3.3. Thermodynamic Analysis of Inclusion Formation in Steel

The precipitation of each sample during the cooling process was calculated by FactSage8.1, a thermodynamic calculation software; the result is shown in Figure 9. The inclusions without Ce were primarily the Al_2O_3-SiO_2-MgO and MnS. When the Ce content was 52 ppm, a large amount of dispersed Ce_2O_2S was formed at 1600 °C. Ce_2O_2S acts as a heterogeneous nucleation core during the cooling process of molten steel, which facilitates MnS precipitation on its surface and avoids the formation of long-striped MnS. As can be seen in Figure 9c, when the Ce content was 139 ppm, CeS began to form at high temperatures. During the cooling process, CeS was first converted to Ce_3S_4 and finally to Ce_2S_3 near the liquidus temperature. With a further increase in the Ce content by 171 ppm, the Ce_2O_2S content slightly changed, the Ce_2S_3 content increased gradually, and the amount of MnS precipitates decreased gradually, as shown in Figure 9d. As shown in Figure 9e,f, when the Ce content was 256 and 340 ppm, MnS precipitation did not occur at all, and the stable sulfide phase of solidified Ce was changed from single Ce_2S_3 to Ce_2S_3, Ce_3S_4, and CeS and Ce_3S_4, respectively, which is consistent with the analysis results of the abovementioned inclusion morphology. Through the above analysis, we confirm that Ce plays a role in mitigating the formation of long-striped MnS through two ways: (1) Ce induces the precipitation of MnS on the surface of sulfur oxides, and (2) the total amount of MnS precipitates is reduced upon S consumption.

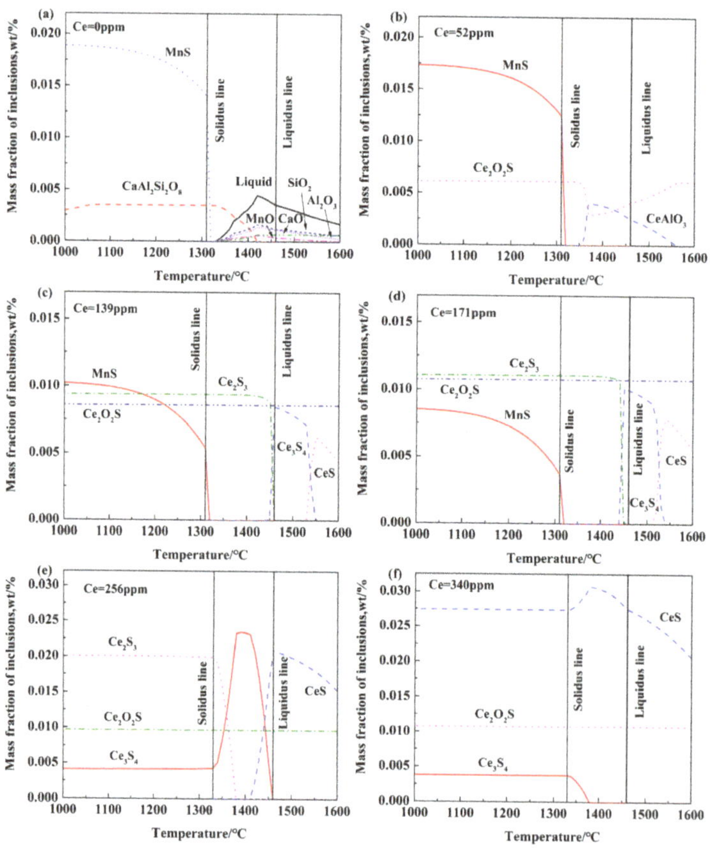

Figure 9. Evolution of inclusions during solidification with different Ce contents: (**a**) 0 ppm, (**b**) 52 ppm, (**c**) 139 ppm, (**d**) 171 ppm, (**e**) 256 ppm, (**f**) 340 ppm.

3.4. Evolution Mechanism of Inclusions in Heavy Rail Steel after Adding Ce

Based on the thermodynamic calculation and the experimental results, the evolution mechanism of the inclusions in U75V heavy rail steel after Ce addition was investigated, as illustrated in Figure 10. According to the composition and morphology analysis of the inclusions, mass Ce_2O_2S phases exist in the inclusions upon Ce addition, which mainly have two sources. (1) The entry of Ce into molten steel and its combination with O and S in the molten steel, which can be expressed as Equation (1) [27], and (2) the modification of the SiO_2-Al_2O_3-CaO inclusion upon Ce addition. When Ce is not added to molten steel, the inclusions at high temperature mainly comprise the SiO_2-Al_2O_3-CaO system. These inclusions are generally large and irregular in shape, while MnS is precipitated during solidification [28]. Therefore, Ce added to the molten steel first diffuses to the interface between the oxide inclusions and molten steel in the form of atoms, followed by their conversion to the ionic form and attachment to the surface of the oxide inclusions. Second, Ce ions react with the active sites on the oxide surface, and reaction products $CeAlO_3$ and Ce_2O_3 adhere to the surface of the original inclusion in the form of a liquid film and grow. As the reaction proceeds, Ce and O in the molten steel continue to diffuse into the interface; the reaction proceeds smoothly, with an increase in the thickness of the liquid film. Once the liquid film reaches a specific thickness, it solidifies and aggregates into spheres according to the principle of minimum surface energy [26]. Thus, a composite inclusion with Si-Al-Ca-O as the core and Ce-Si-Al-Ca oxide as the surface is formed. Al and Si in the inner layer also diffuses to the outer layer with further reaction. Ce ions continue to diffuse to the inner layer, and the required diffusion driving force increases accordingly. At this stage, the ion exchange process of Ce and Al, Si and Ca slows down. Till the end of the diffusion process of internal and external ions, Ce completes the modification of Al-Si-Ca oxide in steel, which can be expressed as Equation (2) [29]. At this stage, the content of Ce in molten steel is sufficient to allow the reaction between Ce oxide S in molten steel, generating oxygen sulfide, which can be expressed as Equation (3) [30].

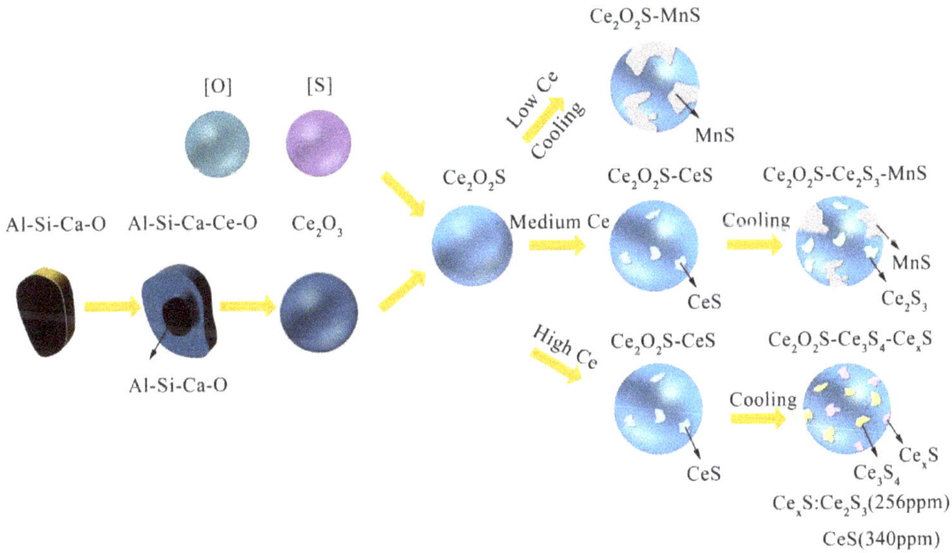

Figure 10. The evolution mechanism of main inclusions in the present study.

References

1. Luu, W.; Wu, J. Effects of sulfide inclusion on hydrogen transport in steels. *Mater. Lett.* **1995**, *24*, 175–179. [CrossRef]
2. Garet, M.; Brass, A.M.; Haut, C.; Guttierez-Solana, F. Hydrogen trapping on non metallic inclusions in Cr-Mo low alloy steels. *Corros. Sci.* **1998**, *40*, 1073–1086. [CrossRef]
3. Domizzi, G.; Anteri, G.; Ovejero-Garcıa, J. Influence of sulphur content and inclusion distribution on the hydrogen induced blister cracking in pressure vessel and pipeline steels. *Corros. Sci.* **2001**, *43*, 325–339. [CrossRef]
4. Wu, X.; Wu, L.; Xie, J.; Shen, P.; Fu, J. Modification of sulfide by Te in Y_1Cr_{13} free-cutting stainless steel. *Metall. Res. Technol.* **2020**, *117*, 107. [CrossRef]
5. Brimacombe, J.K.; Sorimachi, K. Crack formation in the continuous casting of steel. *Metall. Trans. B* **1977**, *8*, 489–505. [CrossRef]
6. Oikawa, K.; Ishida, K.; Nishizawa, T. Effect of titanium addition on the formation and distribution of MnS inclusions in steel during solidification. *ISIJ Int.* **1997**, *37*, 332–338. [CrossRef]
7. Xie, J.; Zhang, D.; Yang, Q.; An, J.; Huang, Z.; Fu, J. Exploration of morphology evolution of the inclusions in Mg-treated 16MnCrS5 steel. *Ironmak. Steelmak.* **2018**, *46*, 564–573. [CrossRef]
8. Lu, J.; Cheng, G.; Chen, L.; Xiong, G.; Wang, L. Distribution and morphology of MnS inclusions in resulfurized non-quenched and tempered steel with Zr addition. *ISIJ Int.* **2018**, *58*, 1307–1315. [CrossRef]
9. Blais, C.; L'Espérance, G.; LeHuy, H.; Forget, C. Development of an integrated method for fully characterizing multiphase inclusions and its application to calcium-treated steels. *Mater. Charact.* **1997**, *38*, 25–37. [CrossRef]
10. Chen, L. Study on the effects of Ti micro-addition on the characteristics of MnS inclusions in rail steel. *Ironmak. Steelmak.* **2019**, *46*, 508–512. [CrossRef]
11. Ji, S.; Zhang, L.; Wang, X. Effect of Magnesium on Inclusions in a High Sulfur Steel. *Metall. Mater. Trans. B* **2022**, *53*, 848–863. [CrossRef]
12. Shen, P.; Fu, J. Morphology study on inclusion modifications using Mg–Ca treatment in resulfurized special steel. *Materials* **2019**, *12*, 197. [CrossRef] [PubMed]
13. Guo, Y.; He, S.; Chen, G.; Wang, Q. Thermodynamics of complex sulfide inclusion formation in Ca-treated Al-killed structural steel. *Metall. Mater. Trans. B* **2016**, *47*, 2549–2557. [CrossRef]
14. Wang, H.; Xiong, L.; Zhang, L.; Wang, Y.; Shu, Y.; Zhou, Y. Investigation of RE-OS-As inclusions in high carbon steels. *Metall. Mater. Trans. B* **2017**, *48*, 2849–2858. [CrossRef]
15. Adabavazeh, Z.; Hwang, W.; Su, Y. Effect of adding cerium on microstructure and morphology of Ce-based inclusions formed in low-carbon steel. *Sci. Rep.* **2017**, *7*, 46503. [CrossRef] [PubMed]
16. Wang, H.; Bao, Y.; Zhao, M.; Wang, M.; Yuan, X.; Gao, S. Effect of Ce on the cleanliness, microstructure and mechanical properties of high strength low alloy steel Q690E in industrial production process. *Int. J. Miner. Metall. Mater.* **2019**, *26*, 1372–1384. [CrossRef]
17. Wang, X.; Li, G.; Liu, Y.; Wang, F.; Wang, Q. Cerium Addition Effect on Modification of Inclusions, Primary Carbides and Microstructure Refinement of H13 Die Steel. *ISIJ Int.* **2021**, *61*, 1850–1859. [CrossRef]
18. Torkamani, H.; Raygan, S.; Mateo, C.G.; Rassizadehghani, J.; Vivas, J.; Palizdar, Y.; San-Martin, D. The influence of La and Ce addition on inclusion modification in cast niobium microalloyed steels. *Metals* **2017**, *7*, 377. [CrossRef]
19. Ren, Q.; Zhang, L. Effect of cerium content on inclusions in an ultra-low-carbon aluminum-killed steel. *Metall. Mater. Trans. B* **2020**, *51*, 589–600. [CrossRef]
20. Liu, Z.; Song, B.; Yang, Z.; Cui, X.; Li, L.; Wang, L.; Song, Z. Effect of Cerium Content on the Evolution of Inclusions and Formation of Acicular Ferrite in Ti-Mg-Killed EH36 Steel. *Metals* **2020**, *10*, 863. [CrossRef]
21. Liu, H.; Fu, P.; Liu, H.; Sun, C.; Sun, M.; Li, D. A novel large cross-section quenching and tempering mold steel matching excellent strength–hardness–toughness properties. *Mat. Sci. Eng. A* **2018**, *737*, 274–285. [CrossRef]
22. Liu, Y.; Wang, L.; Chou, K. Effect of cerium on the cleanliness of spring steel used in fastener of high-speed railway. *J. Rare Earths* **2014**, *32*, 759–766. [CrossRef]
23. Gao, S.; Wang, M.; Guo, J.; Wang, H.; Zhi, J.; Bao, Y. Characterization Transformation of Inclusions Using Rare Earth Ce Treatment on Al-Killed Titanium Alloyed Interstitial Free Steel. *Steel Res. Int.* **2019**, *90*, 1900194. [CrossRef]
24. Luo, S.; Shen, Z.; Yu, Z.; Wang, W.; Zhu, M. Effect of Ce Addition on Inclusions and Grain Structure in Gear Steel 20CrNiMo. *Steel Res. Int.* **2021**, *92*, 2000394. [CrossRef]
25. Wang, Y.; Li, C.; Wang, L.; Xiong, X.; Chen, L.; Zhuang, C. Modification of Alumina Inclusions in SWRS82B Steel by Adding Rare Earth Cerium. *Metals* **2020**, *10*, 1696. [CrossRef]
26. Zhang, J.; Yu, Y.; Wang, S.; Hou, Y.; Yin, S. Effects of Rare Earth on Mechanical Properties of LZ50 Axle Steels and Its Formation Mechanism. *High Temp. Mater. Process* **2018**, *37*, 509–519. [CrossRef]
27. Cai, G.; Pang, Y.; Huang, Y.; Misra, R.D.K. Roles of Inclusion, Texture and Grain Boundary in Corrosion Resistance of Low-Nickel Austenite Stainless Steel Containing Ce. *ISIJ Int.* **2019**, *59*, 2302–2310. [CrossRef]
28. Shen, P.; Yang, Q.; Zhang, D.; Yang, S.; Fu, J. The effect of tellurium on the formation of MnTe-MnS composite inclusions in non-quenched and tempered steel. *Metals* **2018**, *8*, 639. [CrossRef]
29. Li, X.; Jiang, Z.; Geng, X.; Chen, M.; Peng, L. Evolution mechanism of inclusions in H13 steel with rare earth magnesium alloy addition. *ISIJ Int.* **2019**, *59*, 1552–1561. [CrossRef]

30. Gong, W.; Wang, P.; Zhang, L.; Jiang, Z. Effects of Ce on Microstructure and Mechanical Properties of LDX2101 Duplex Stainless Steel. *Metals* **2020**, *10*, 1233. [CrossRef]
31. Li, X.; Jiang, Z.; Geng, X.; Chen, M.; Cui, S. Effect of Rare Earth–Magnesium Alloy on Inclusion Evolution in Industrial Production of Die Steel. *Steel Res. Int.* **2019**, *90*, 1900103. [CrossRef]

metals

MDPI

Article

Source and Transformation of MgO-Based Inclusions in Si-Mn-Killed Steel with Lime-Silicate Slag

Jiaqi Zhao [1,2], Jianhua Chu [3], Xin Liu [1], Min Wang [1], Xiaofeng Cai [2], Han Ma [2] and Yanping Bao [1,*]

1 State Key Lab of Advanced Metallurgy, University of Science & Technology Beijing, Beijing 100083, China
2 Institute of Research of Iron & Steel, Zhangjiagang 215600, China
3 School of Metallurgical Engineering, Anhui University of Technology, Maanshan 243002, China
* Correspondence: baoyp@ustb.edu.cn

Abstract: The origin, evolution, and formation mechanism of MgO-based inclusions in Si-Mn-killed steel were studied in industrial trials with systematical samplings of the refining ladle, casting tundish, and as-cast bloom. In the present study, there were large numbers of MgO-based non-metallic inclusions, which started to form in the LF final process, and the MgO content in the lime-silicate slag increases from LF to VD process. The reason for the formation of MgO-based inclusions in refining process was analyzed using FactSage8.1 software. It was found that MgO-based inclusions were caused by the violent reaction between the slag and steel and the serious erosion of MgO-C refractory. The MgO solubility decreased in the lime-silicate slag and precipitated the periclase phase with basicity increasing. The solubility of MgO increased with an increase in the temperature. Measures were taken to optimize the refining process based on the above result. By increasing the slag basicity and increasing the content of MgO in the slag, erosion of the MgO-C refractory was reduced and the number of MgO-based non-metallic inclusions decreased from 0.2 to 0.04 per square millimeter.

Keywords: MgO saturation; industrial trial; non-metallic inclusions; thermodynamic calculation

Citation: Zhao, J.; Chu, J.; Liu, X.; Wang, M.; Cai, X.; Ma, H.; Bao, Y. Source and Transformation of MgO-Based Inclusions in Si-Mn-Killed Steel with Lime-Silicate Slag. *Metals* **2022**, *12*, 1323. https://doi.org/10.3390/met12081323

Academic Editor: Alexander Ivanovich Zaitsev

Received: 7 June 2022
Accepted: 1 August 2022
Published: 7 August 2022

1. Introduction

Inclusions which are higher in hardness and melting point are harmful to high-strength steel rods because inclusions often act as breakage initiation sites when subjected to cyclic stress. It would be much more desirable for the remained inclusions to be small in size and to demonstrate improved deformability, which is affected by the melting points of the inclusions themselves [1,2]. Therefore, inclusions, such as alumina or magnesia which are higher in melting point and hardness, should be avoided. For these reasons, Si-Mn complex deoxidation combined with low-basicity silicate slag refining is generally usedfor such steels [3–5].

The common type of non-metallic inclusions in these steels is $CaO\text{-}SiO_2\text{-}Al_2O_3$, usually together with some MgO [4]. Many researchers studied the effect of the Al_2O_3 and SiO_2 content on inclusions in Si-Mn-killed steel with lime-silicate slag, through either experiments or thermodynamic predictions [5–15]. Li Y et al. [5] studied the influence of a MgO refractory on inclusions and found thatinclusions containing MgO originated from the reduction of MgO in refractory or slag. Liu et al. [6] found that C in the MgO-C refractory also reduced MgO in the refractory and supplied Mg to the steel melt. Harada et al. [7] observed that the Al in the steel melt reduced the MgO in the refractory and Mg was dissolved into the steel melt. According to this mechanism, MgO in the MgO-C refractory supplies Mg to the steel melt owing to the reduction in Al in the steel. Chen L et al. [8] found that the refractory/steel interaction under vacuum conditions has a far more significant contribution to the increase of the $[Al]_S$ (acid-soluble Al) content in steel compared with the slag/steel interaction, which further leads to the generation of Al_2O_3-rich inclusions. Wang K et al. [9] studied the formation mechanism of CaO-containing inclusions in tire cord steel. Park J et al. [10], Chen S et al. [11] and Liu N et al. [12] also studied the effect of refining slag basicity on non-metallic inclusions in steel. Additionally, Zhang L et al. [13–15] thought that kinetic factors were also of great importance in the formation of non-metallic inclusions. Moreover, many researchers [16–18] have studied the solubility of MgO in the CaO-SiO_2-FetO-MgO system. The solubility of MgO in low-basicity lime-silicate slag decreases with increasing the slag basicity (CaO/SiO_2). However, research and control on the silicate inclusions surrounding bulk MgO are still limited.

In the current study, industrial experiments were performed to investigate the composition and morphology of inclusions in the whole process including the steelmaking process. Based on the experiment results and thermodynamic calculation, the formation mechanism of MgO-based non-metallic inclusions is put forward.

2. Experiments

2.1. Experiment and Samplings

An industrial trial was carried out in a steel plant for Si-Mn-killed steel with the chemical composition listed in Table 1. The steelmaking process featured BOF (basic oxygen furnace steelmaking),-LF (ladle refining),-VD (vacuum degassing),-TD (soft blowing-tundish),-CC (continuous casting, 150 mm × 150 mm section). During tapping of the BOF, FeSi and Mn were used as deoxidizers and alloying elements. Besides, Cr–Fe alloy and refining slag material were also added. In the LF process, Si–Fe and Mn were employed for composition modification. Soft blowing was couducted for about 25 min to remove inclusions after VD degassing. Finally, the molten steel was sent to continuous casting.

Table 1. Chemical composition of the experimental steel, wt%.

C	Si	Mn	Cr	P	S	Als
0.50–0.55	0.30–0.35	0.40–0.50	0.20–0.30	<0.014	<0.012	<0.0035

To investigate the source and transformation of MgO-based inclusions, steel samples and slag samples were taken with samplers (diameter of 30 mm, thickness of 10 mm): three from LF refining (start, after alloying, end), two from VD refining (start, end), one from soft blowing, one from the tundish, and one from billet casting, as shown in Figure 1.

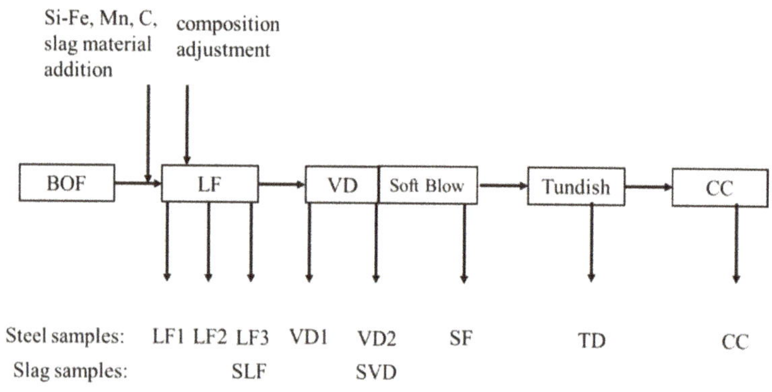

Figure 1. Process of industrial trials and sampling locations.

2.2. Analysis of Samples

These samples were polished for characterization of the inclusions using scanning electron microscopy (SEM). The composition of inclusions was analyzed by an energy dispersive X-ray spectrometer (EDS) attached to the SEM. The slag composition shown in Table 2 was determined using an X-ray fluorescence spectrometer (XRF) with a 5 pct relative standard deviation.

Table 2. Chemical composition of the refining slag, wt%.

Sampling Location	CaO	SiO_2	Al_2O_3	MgO	T.Fe+MnO
SLF	39–44	37–43	5–10	3–7	≤4
SVD	38–42	37–45	4–8	8–13	≤5

3. Results and Discussion

3.1. Evolution of Non-Metallic Inclusions in Steelmaking Process

Inclusions were analyzed using an SEM equipped with an EDS in each metallographic sample taken from different stages, as shown in Figure 2. The types of non-metallic inclusions wre CaO-Al_2O_3-SiO_2 and Al_2O_3-SiO_2-MnO at the early stage of the LF process, but CaO-Al_2O_3-SiO_2-MgO was found in the VD process. During the later stage of the LF process, there were more MgO inclusions, which were irregular and wrapped by CaO-Al_2O_3-SiO_2 inclusions. Those inclusions were not completely removed in the process of soft stirring and tundish pouring, and they eventually remained in the billet. The variation in the average mass fraction of MgO and Al_2O_3 in inclusions during the refining process is shown in Figure 3. The average Al_2O_3 content of oxide inclusions gradually decrease from 43% to 15% in the refining process. The average MgO content of oxide inclusions increase from 7% to 15%. The average MgO content in inclusions increased rapidly from the late refining stage to the VD stage.

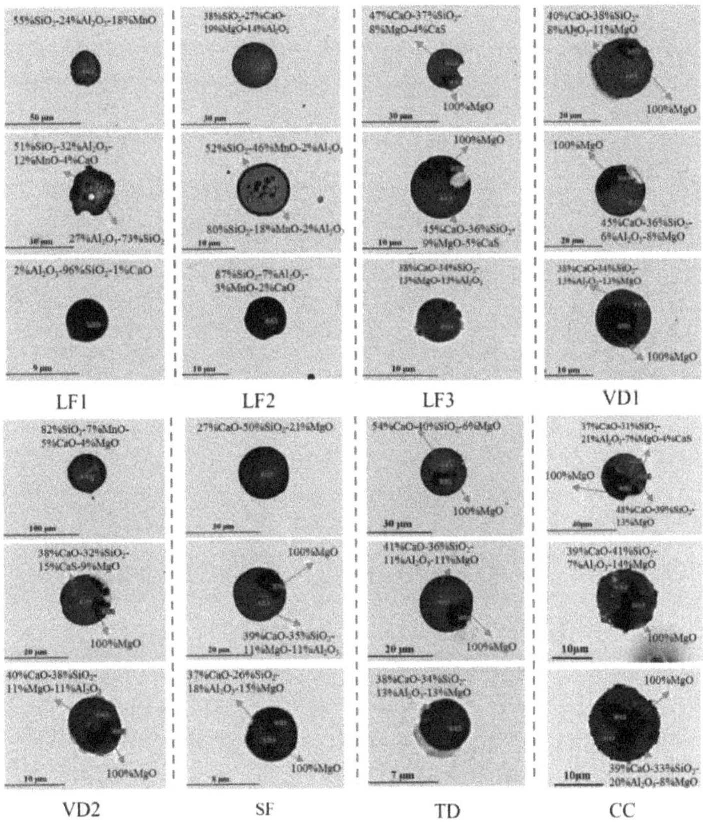

Figure 2. Morphology and composition of typical inclusions in steel.

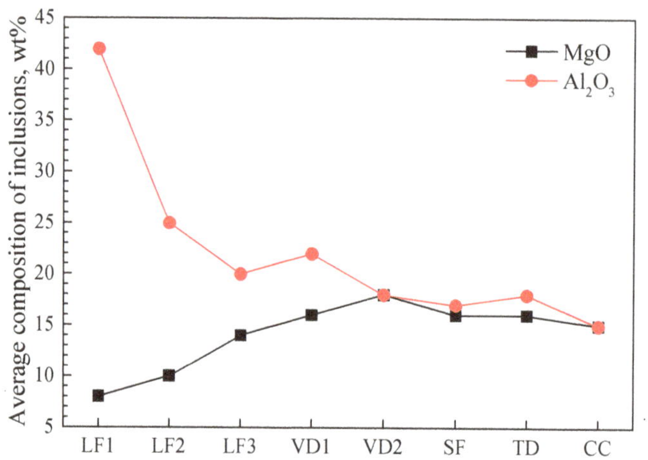

Figure 3. Average mass fraction of Al_2O_3 and MgO in $CaO-Al_2O_3-MgO-SiO_2$ inclusions.

3.2. Analysis of the MgO-C Refractory after Continuous Casting

By taking samples of refractory materials at the slag line of the ladle furnace after smelting, the longitudinal section of the refractory contacting the molten steel was analyzed with a scanning electron microscope, as shown in Figure 4. From inside to outside, it can be divided into three layers, including the complete reaction layer, transition layer and matrix. Almost all C in the reaction layer and transition layer of the refractory was oxidized, and the reaction layer infiltrated obvious Ca-Si-Al-O inclusions and Fe droplets. Further analysis of the micro area in the reaction layer is shown in Figure 5. There was a large amount of massive MgO in the refractory, which was wrapped by infiltrated Ca-Si-Al-O inclusions.

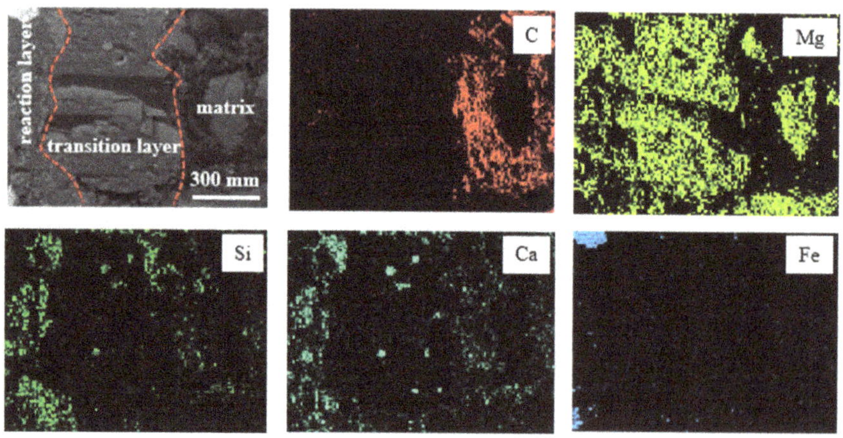

Figure 4. Longitudinal sectional microstructure and element mapping of the ladle furnace slag line MgO-C bricks.

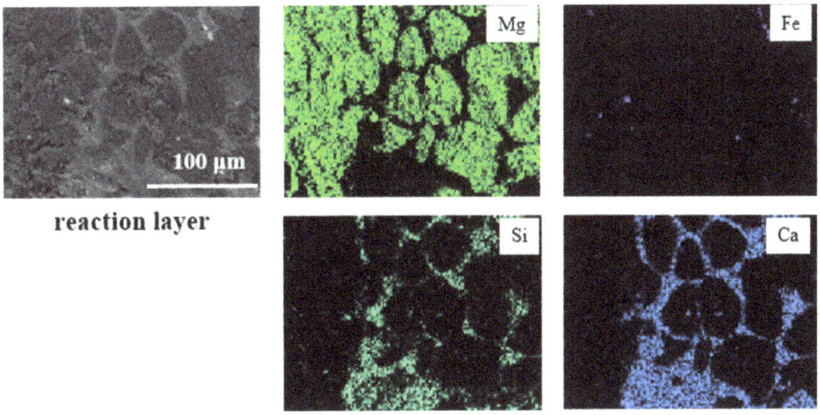

Figure 5. Microstructure and element mapping of the reaction layer in the slag line MgO-C bricks.

3.3. Source of MgO-Based Non-Metallic Inclusions

Inclusions were analyzed using an SEM equipped with an EDS in the billet as shown in Figure 6. It was found that the inclusions in steel were CaO-Al_2O_3-SiO_2 and Al_2O_3-SiO_2-MnO, and that there were a large number of composite inclusions with massive MgO wrapped by silicate inclusions. The number of MgO-based inclusions in the billet is shown in Table 3. The size distribution of all inclusions and MgO-based inclusions in billet is shown in Figure 7, and the morphology and composition of typical inclusions are shown in Figure 8. It can be seen that the light gray part of the inclusion corresponds to the CaO-SiO_2-Al_2O_3 phase, and the black area depicts the MgO phase. The shape of the MgO phase is irregular and has edges and corners.

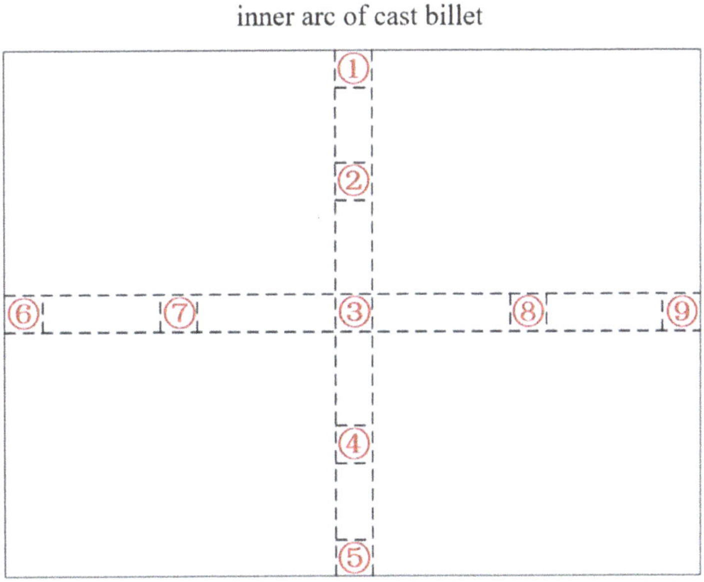

Figure 6. Sampling plan.

Table 3. Non-metallic inclusions in billet before improvement.

NO.	Number of All Types of Inclusions	MgO-Based Non-Metallic Inclusions	
		Number	Density, $/mm^2$
1	177	28	0.32
2	204	26	0.29
3	272	5	0.06
4	148	14	0.16
5	134	9	0.10
6	333	15	0.17
7	298	22	0.25
8	332	23	0.26
9	207	15	0.17
Average			0.20

The MgO inclusions probably came from the refining slag or corrosion of the MgO-C refractory. Refining slag with a low basicity has strong corrosiveness. The MgO content in the refining slag increased from the LF to the VD process, as shown in Table 2. The thermodynamic FactSage8.1 software with FToxid databases, SafeNet Inc., Belcamp, MD, USA (for oxides) was used to simulate the precipitation of inclusions during solidification. Combined with the statistical data of the inclusions mentioned above, the average composition of inclusions in the steel was $40CaO\text{-}30SiO_2\text{-}15Al_2O_3\text{-}15MgO$ in the billet. At the molten steel temperature, there are precipitations of MgO from liquid inclusions in the $40CaO\text{-}30SiO_2\text{-}15Al_2O_3\text{-}15MgO$ system, as shown in Figure 9.

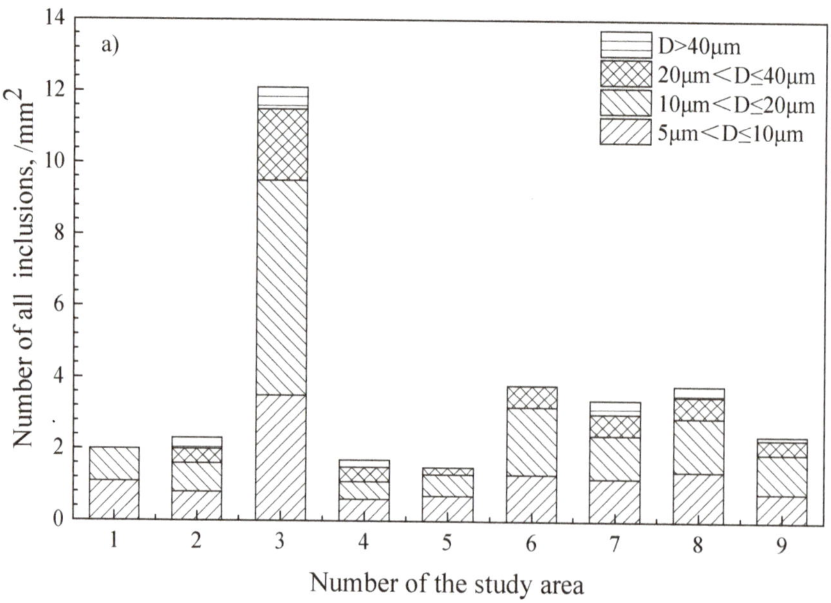

Figure 7. *Cont.*

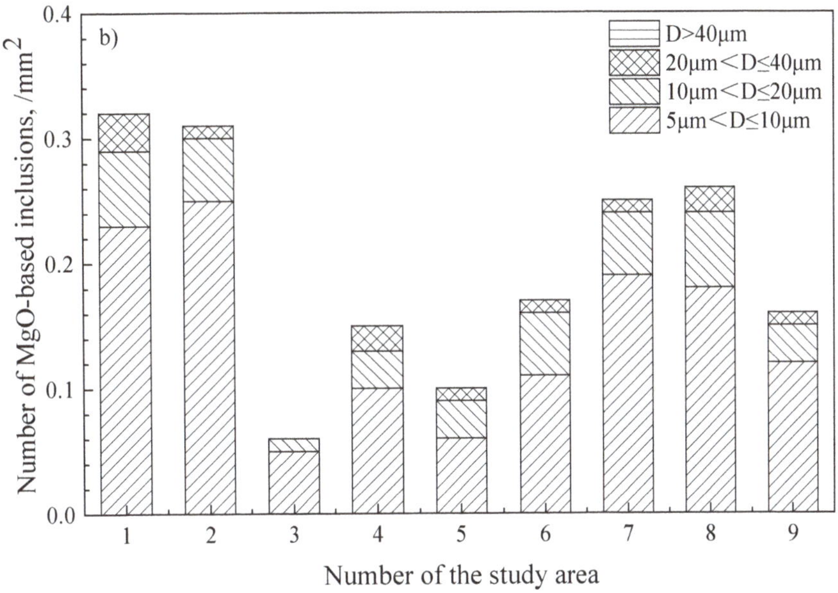

Figure 7. Size distribution of inclusions in the billet. (**a**) all types of inclusions; (**b**) MgO-based inclusions.

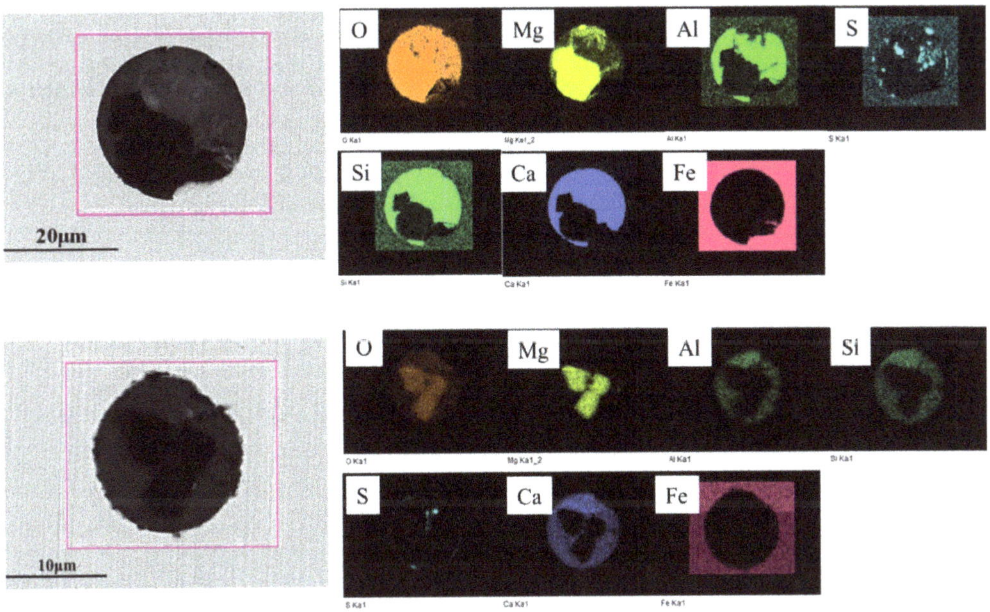

Figure 8. Morphology and composition of typical inclusions in billet.

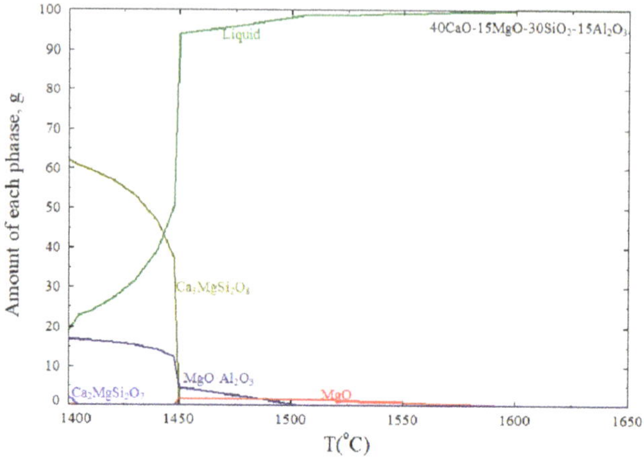

Figure 9. Precipitation of inclusions during solidification.

3.4. Thermodynamic Considerations of MgO Saturation in Slag

MgO-unsaturated secondary metallurgical slag can cause MgO to dissolve from a refractory brick into slag. The MgO solubility in slag is controlled by saturation phases that include periclase (MgO), spinel ($MgO{\cdot}Al_2O_3$) or forsterite ($2MgO{\cdot}SiO_2$) in the quaternary CaO-SiO_2-Al_2O_3-MgO system. The thermodynamic FactSage8.1 software was used to obtain the MgO saturation concentration under different conditions by the isothermal stability diagrams (ISDs) [19].

Figure 10 shows the effect of the Al_2O_3 content on MgO solubility with 25~40% SiO_2 contents at 1873 K. The MgO solubility was about 2~34% with 0~10% Al_2O_3 content in the quaternary CaO-SiO_2-Al_2O_3-MgO system. With an increase in the Al_2O_3 content, the MgO solubility in the MgO-saturated phases periclase (MgO), forsterite (Mg_2SiO_4) and dual saturated (MgO and Ca_2SiO_4) increased, and thereafter, decreased in the MgO-saturated phase spinel ($MgO{\cdot}Al_2O_3$). Additionally, the SiO_2 content had a significant effect on the MgO solubility in the lime-silicate slag. The MgO solubility increased with an increase to less than 40% SiO_2 content.

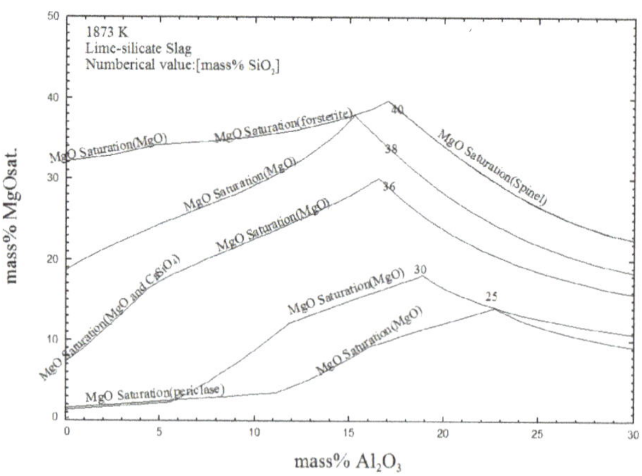

Figure 10. Effect of the Al_2O_3 content on MgO solubility in lime-silicate slag at 1873 K.

Figure 11 displays the effect of the Al_2O_3 content at different basicities (CaO/SiO_2) on MgO solubility at 1873 K. It is notable that slags with higher basicity were saturated with less MgO. Typical lime-silicate slag had less than 10% Al_2O_3 content and 0.8~2.0 basicity, therefore, the solubility of MgO was 1~29%. This conclusion was consistent with previous researches [7–9] considering the solubility of MgO in the $CaO-SiO_2-Fe_tO-MgO$ system.

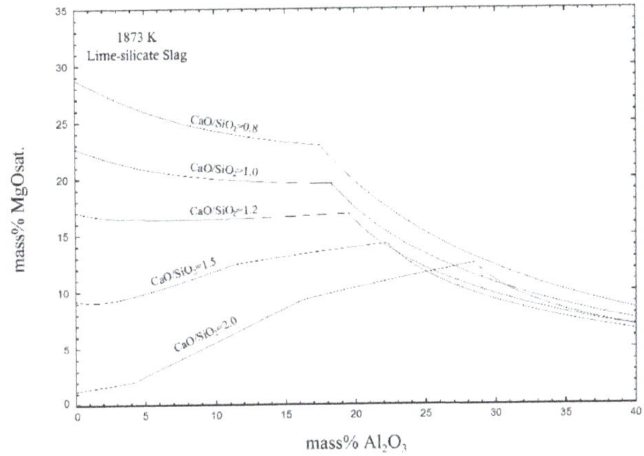

Figure 11. Effect of basicity (CaO/SiO_2) on MgO solubility with a low basicity of the slag at 1873 K.

Figure 12 shows the effect of the Al_2O_3 content at different temperature on the MgO solubility with 40% SiO_2 content. More MgO was saturated in the quaternary $CaO-SiO_2-Al_2O_3-MgO$ system at a higher temperature.

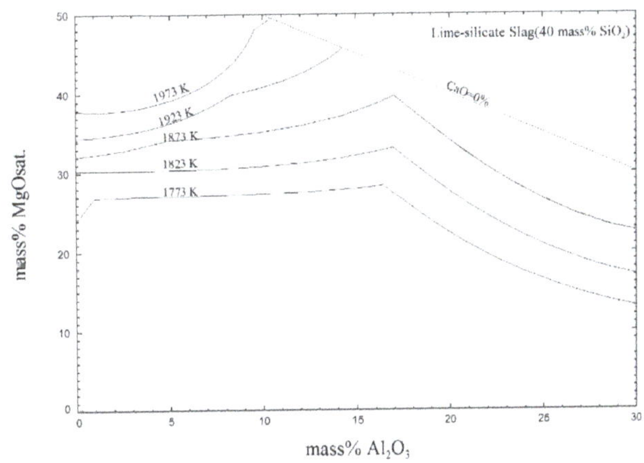

Figure 12. Effect of the Al_2O_3 content on MgO solubility in slag with 40% SiO_2 at different temperatures.

3.5. Evolution Mechanism of MgO-Based Non-Metallic Inclusions

The main components of refractory materials are MgO and C. Because the MgO content in the slag is not saturated, the basicity of the refining slag is low, and the oxidation property is high, the C in the refractory is oxidized, and MgO gradually enters into the slag, resulting in the gradual increase of the MgO content in the slag and the gradual erosion of the refractory [20]. At the later stage of smelting, the impurities in the steel are mainly $CaO-SiO_2-Al_2O_3$ low melting point liquid inclusions, the C in the refractory is oxidized, and the CaO-SiO2-Al_2O_3 inclusions in the steel enter the surface cracks of the refractory. As Liu [6] found, C in the MgO-C refractory also reduced MgO in the refractory in Al-killed steel. The main difference was that a MgO-based inclusion was not formed by Mg from the refractory in the steel. With the scouring of the refractory by the bottom blowing of the ladle and the flow of the steel, the massive MgO flakes into the steel, or the $CaO-SiO_2$-Al2O3 wraps the massive MgO inclusions and washes into the steel. The detailed mechanism process is illustrated in Figure 13. The evolution mechanism of MgO-based inclusions is shown in Figure 14.

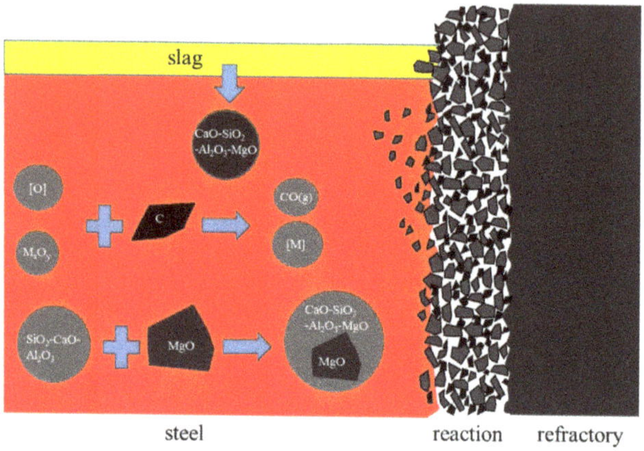

Figure 13. Illustration diagram of steel-slag-refractory interaction mechanism.

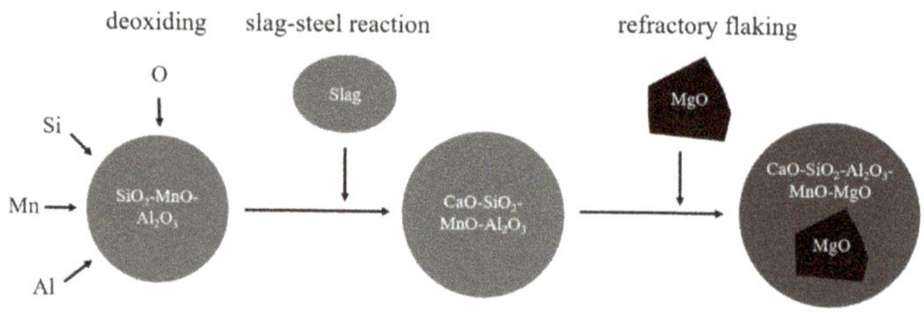

Figure 14. Schematic diagram of MgO-based inclusion evolution mechanism.

3.6. Improvements in Industrial Trial

From the analysis above, the corrosion of the refractory can be reduced by increasing the basicity and MgO content in the refining slag. Hence improvements were made in the industrial trial. We redesigned the refining slag, specifically by: (1) appropriately increasing the amount of lime in the LF refining process, increasing the amount of lime by 50–100kg, controlling the basicity of slag to 1.5–2.0 and reducing the saturation solubility of MgO in refining slag; (2) increasing the content of MgO in the refining slag, making the content of MgO in the refining slag reach saturation state and reducing the corrosion of the refractory. The main components of the refining slag before and after improvement are shown in Table 4.

Table 4. Components of the refining slag in LF before and after improvement, wt%.

Different Process	CaO	SiO_2	Al_2O_3	MgO	T.Fe+MnO
before improvement	39–44	37–43	5–10	3–7	$\leq$5.0
after improvement	45–52	27–33	3–8	8–14	$\leq$1.5

Inclusions were analyzed using an SEM equipped with an EDS in the billet, as shown in Figure 15. After improvement, the samples of the continuous casting billet were analyzed. The number and density of inclusions are shown in Table 5. The morphology and composition of typical inclusions in billet are shown in Figure 16. The inclusions in the steel were CaO-Al_2O_3-SiO_2 and Al_2O_3-SiO_2-MnO, some of which also contained a minor amount of MgO, but the composition of MgO was relatively uniform in distribution, and the number of composite inclusions wrapped with MgO decreased greatly.

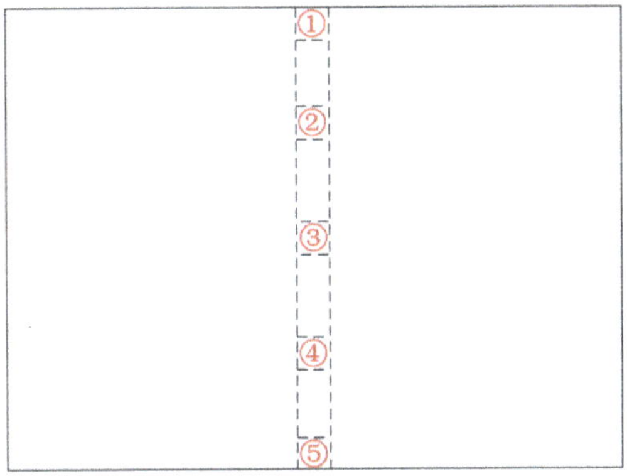

Figure 15. Sampling plan.

Table 5. Non-metallic inclusions in the billet after improvement.

NO.	Number of all Types of Inclusions	MgO-Based Non-Metallic Inclusions	
		Number	Density, /mm^2
1	168	2	0.03
2	243	5	0.06
3	261	5	0.06
4	177	2	0.02
5	244	4	0.05
Average			0.04

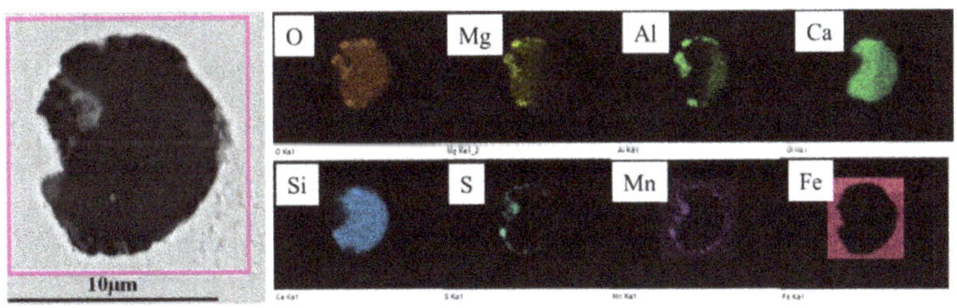

Figure 16. Morphology and composition of typical inclusions in billet.

Before improvement the average density of oxide inclusions containing MgO was 0.59/mm^2, while that of massive MgO inclusions was 0.2/mm^2. After improvement, the average density of oxide inclusions containing MgO was 0.45/mm^2, and the density of MgO-based oxide inclusions was reduced to 0.04/mm^2 in Figure 17. The proportion of MgO-based oxide inclusions decreased from 6.8% to 1.7%. The objective of controlling MgO-based oxide inclusions was achieved by optimizing the refining process.

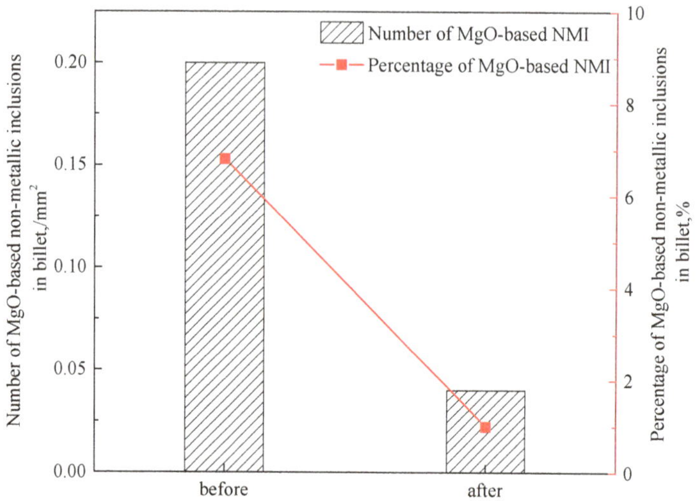

Figure 17. The number of MgO-based oxide inclusions in the billet.

4. Conclusions

In the present study, industrial trials and thermodynamic calculations were performed to investigate the solubility of MgO in the quaternary CaO-SiO_2-Al_2O_3-MgO system in secondary metallurgical slag. The following conclusions were obtained:

1. By using an SEM equipped with an EDS, it was found that there were a large number of MgO-based non-metallic inclusions, which started to form in the LF final process. The content of MgO in the lime-silicate slag increased from the LF process to VD process, which may have been caused by the erosion of MgO-C refractory.
2. The solubility of MgO in low basicity lime-silicate slag decreased with an increase in the slag basicity (CaO/SiO_2). In the lime-silicate slag, the solubility of MgO increases with an increase in the temperature. The solubility of MgO was 1~29% in typical lime-silicate slag with 0~10% Al_2O_3 and 0.8~2.0 slag basicity (CaO/SiO_2) at 1873 K.
3. With the oxidation of C in the refractory and the scouring of the refractory by the bottom blowing of the ladle and the flow of moten steel, the massive MgO flakes into the steel, or the CaO-SiO_2-Al_2O_3 wraps the massive MgO inclusions and washes into the steel.
4. By increasing the slag basicity and increasing the content of MgO, erosion of MgO-C refractory was reduced and number of MgO-based non-metallic inclusions was decreased from 0.2 to 0.04 per square millimeter. The objective of controlling MgO-based oxide inclusions was achieved by optimizing the refining process.

Author Contributions: Conceptualization, J.Z. and J.C.; methodology, J.Z.; software, X.C.; validation, J.Z., J.C. and M.W.; formal analysis, J.C.; investigation, J.Z.; resources, J.Z.; data curation, J.Z.; writing—original draft preparation, J.Z.; writing—review and editing, X.L.; visualization, X.C; supervision, M.W., H.M. and Y.B.; project administration, H.M.; funding acquisition, J.Z. All authors have read and agreed to the published version of the manuscript.

Funding: This work was supported by the National Natural Science Foundation of China (Grant No. 51774031).

Data Availability Statement: The data presented in this study are available on request from the corresponding author.

Conflicts of Interest: No potential conflict of interest was reported by the authors.

References

1. Chen, C.; Jiang, Z.; Li, Y.; Sun, M.; Wang, Q.; Chen, K.; Li, H. State of the Art in the Control of Inclusions in Spring Steel for Automobile—A Review. *ISIJ Int.* **2020**, *60*, 617–627. [CrossRef]
2. Bernard, G.; Ribound, P.; Urbain, G. Oxide inclusions plasticity. *Rev. Metall. Cah. Inf. Tec.* **1981**, *78*, 421. [CrossRef]
3. Kang, Y.-B.; Lee, H.-G. Inclusions Chemistry for Mn/Si Deoxidized Steels: Thermodynamic Predictions and Experimental Confirmations. *ISIJ Int.* **2004**, *44*, 1006–1015. [CrossRef]
4. Kirihara, K. Production technology of wire rod for high tensile strength steel cord. *Kobelco Technol. Rev.* **2011**, *30*, 62–65.
5. Li, Y.; Yang, W.; Zhang, L. Formation Mechanism of MgO Containing Inclusions in the Molten Steel Refined in MgO Refractory Crucibles. *Metals* **2020**, *10*, 444. [CrossRef]
6. Liu, C.; Gao, X.; Kim, S.; Ueda, S.; Kitamura, S. Dissolution Behavior of Mg from MgO-C Refractory in Al-killed Molten Steel. *ISIJ Int.* **2018**, *58*, 488–495. [CrossRef]
7. Harada, A.; Miyano, G.; Maruoka, N.; Shibata, H.; Kitamura, S.-Y. Dissolution Behavior of Mg from MgO into Molten Steel Deoxidized by Al. *ISIJ Int.* **2014**, *54*, 2230–2238. [CrossRef]
8. Chen, L.J.; Chen, W.Q.; Hu, Y.; Chen, Z.P.; Xu, Y.T.; Yan, W. Investigation on the Origin of Al_2O_3-Rich Inclusions in Valve Spring Steel under Vacuum Condition. *Steel Res. Int.* **2017**, *88*, 1600376. [CrossRef]
9. Wang, K.; Jiang, M.; Wang, X.; Wang, Y.; Zhao, H.; Cao, Z. Formation Mechanism of CaO-SiO_2-Al_2O_3-(MgO) Inclusions in Si-Mn-Killed Steel with Limited Aluminum Content During the Low Basicity Slag Refining. *Metall. Mater. Trans. B* **2016**, *47*, 282–290. [CrossRef]
10. Park, J.S.; Park, J.H. Effect of slag composition on the concentration of Al_2O_3 in the inclusions in Si-Mn-killed steel. *Metall. Mater. Trans. B* **2014**, *45*, 953–960. [CrossRef]
11. Chen, S.-H.; Jiang, M.; He, X.-F.; Wang, X.-H. Top slag refining for inclusion composition transform control in tire cord steel. *Int. J. Miner. Met. Mater.* **2012**, *19*, 490–498. [CrossRef]
12. Liu, N.; Zhang, L.; Chu, Y.; Ren, Y. Effect of Slag Basicity on Non-metallic Inclusions in a Heavy Rail Steel. In *12th International Symposium on High-Temperature Metallurgical Processing*; Springer International Publishing: Berlin/Heidelberg, Germany, 2022; pp. 513–520.
13. Zhang, Y.; Ren, Y.; Zhang, L. Termodynamics and kinetics for the evolution of non-metallic inclusions in pipeline steel. *La Metall. Ital.* **2019**, *3*, 25–34.
14. Zhang, L.; Ren, Q.; Duan, H.; Ren, Y.; Chen, W.; Cheng, G.; Yang, W.; Sridhar, S. Modelling of non-metallic inclusions in steel. Mineral Processing and Extractive Metallurgy. *Trans. Inst. Min. Metall.* **2020**, *129*, 184–206.
15. Park, J.H.; Zhang, L. Kinetic Modeling of Nonmetallic Inclusions Behavior in Molten Steel: A Review. *Met. Mater. Trans. A* **2020**, *51*, 2453–2482. [CrossRef]
16. Shim, J.D.; Ban-Ya, S. The solubility of magnesia and ferric-ferrous equilibrium in liquid FetO-SiO_2-CaO-MgO slags. *Tetsu Hagané* **1981**, *67*, 1735–1744. [CrossRef]
17. Jung, S.-M.; Min, D.-J.; Rhee, C.-H. Solubility of MgO in New Ironmaking Process-typed Slags. *ISIJ Int.* **2007**, *47*, 1718–1722. [CrossRef]
18. Tayeb, M.A.; Assis, A.N.; Sridhar, S.; Fruehan, R.J. MgO Solubility in Steelmaking Slags. *Met. Mater. Trans. A* **2015**, *46*, 1112–1114. [CrossRef]
19. Bennett, J.; Kwong, K.-S. Thermodynamic studies of MgO saturated EAF slag. *Ironmak. Steelmak.* **2010**, *37*, 529–535. [CrossRef]
20. Zhang, L.; Thomas, B.G. State of the Art in Evaluation and Control of Steel Cleanliness. *ISIJ Int.* **2003**, *43*, 271–291. [CrossRef]

metals

Article

Formation and Removal Mechanism of Nonmetallic Inclusions in 42CrMo4 Steel during the Steelmaking Process

Tong Qiao [1], Guoguang Cheng [1,*], Yu Huang [1], Yao Li [1], Yanling Zhang [1] and Zhanchun Li [2]

1 State Key Laboratory of Advanced Metallurgy, University of Science and Technology Beijing, Beijing 100083, China
2 Jiangsu Yonggang Group Co., Ltd., Zhang Jiagang 215628, China
* Correspondence: chengguoguang@metall.ustb.edu.cn

Abstract: Nonmetallic inclusions are harmful to the quality of 42CrMo4 steel. Therefore, the formation and removal mechanism of inclusions in 42CrMo4 steel during the steelmaking process is investigated by industrial trials. The characteristics of inclusions in specimens were analyzed by scanning electron microscopy and energy dispersive spectroscopy. The main type of inclusions in molten steel in the early stage of ladle furnace (LF) refining is MgO-Al_2O_3 inclusions of irregular shape. CaO begins to appear in MgO-Al_2O_3 inclusions in the middle and late stages of LF. In the vacuum degassing (VD) refining stage, the inclusions in molten steel completely change into low-melting-point CaO-MgO-Al_2O_3 inclusions. The existence of [Mg] in molten steel is the fundamental reason for the formation of a large number of MgO-Al_2O_3 inclusions. Thermodynamic calculation shows that the refractory mainly transfers [Mg] to the liquid steel in the LF refining stage, whereas the slag mainly transfers [Mg] to the liquid steel in the VD refining stage. Kinetic calculation indicates that MgO-Al_2O_3 inclusions could be removed from molten steel faster than low-melting-point CaO-MgO-Al_2O_3 inclusions. The fundamental reason for the different removal behavior of the two types of inclusions is that the interfacial tension between the low-melting-point CaO-MgO-Al_2O_3 inclusions and the liquid steel is 50% lower than that of the MgO-Al_2O_3 inclusions.

Keywords: nonmetallic inclusions; MgO-Al_2O_3; CaO-MgO-Al_2O_3; thermodynamic; kinetic

Citation: Qiao, T.; Cheng, G.; Huang, Y.; Li, Y.; Zhang, Y.; Li, Z. Formation and Removal Mechanism of Nonmetallic Inclusions in 42CrMo4 Steel during the Steelmaking Process. *Metals* **2022**, *12*, 1505. https://doi.org/10.3390/met12091505

Academic Editors: Alexander Ivanovich Zaitsev and Lauri Holappa

Received: 13 August 2022
Accepted: 9 September 2022
Published: 11 September 2022

1. Introduction

The 42CrMo4 type of steel is a medium carbon alloy structural steel that has high strength, hardenability, toughness, creep strength, and endurance strength at high temperatures [1–5]. It is commonly used to prepare pitch and yaw bearings in wind power steel. Because of the harsh working environment, complex load-bearing conditions, and the high cost of lifting and replacing bearings, wind power equipment needs longer service life and higher stability. Inclusions in steel are the key factor affecting fatigue life [6–8]. Therefore, it is important to study the formation and removal mechanism of inclusions in 42CrMo4 steel during smelting to reasonably control inclusions in molten steel and improve the final product quality of 42CrMo4 steel.

Aluminum deoxidizes 42CrMo4 steel, and the deoxidized product is Al_2O_3 [9,10]. However, the main inclusions found in the smelting process are MgO-Al_2O_3 inclusions, which will change into CaO- MgO-Al_2O_3 inclusions along with the smelting process. The presence of [Mg] in liquid steel leads to the transformation of Al_2O_3 inclusions into MgO-Al_2O_3, but there is still controversy about the source of [Mg] in molten steel. Some authors [11–14] have pointed out that slag containing MgO and magnesia refractory reacted with liquid steel and transferred [Mg] to liquid steel together. Increasing [Mg] content in liquid steel promotes inclusion change from simple Al_2O_3 inclusions to MgO-Al_2O_3 inclusions. However, some authors [15,16] have noted that the transfer of [Mg] from slag to molten steel was dominant, and refractory materials had little effect on it. There are also authors [17–19] who have pointed out that the displacement reaction between [Al] in molten steel and MgO in refractories was the main reason for the increase in [Mg] in molten steel, which further determined the formation of MgO-Al_2O_3 inclusions. As for the transformation mechanism of MgO-Al_2O_3 inclusions to CaO-MgO-Al_2O_3 inclusions, a large number of authors [20–24] have reported that [Ca] reduced MgO in MgO-Al_2O_3 inclusions to CaO and eventually formed low-melting-point CaO-MgO-Al_2O_3 inclusions. However, compared with solid inclusions, low-melting-point liquid inclusions are more difficult to remove from molten steel [25–31]. As a result, the main type of inclusions observed in molten steel at the later stage of smelting is low-melting-point liquid inclusions, which will be unfavorable to the final product quality. For 42CrMo4 steel, which uses the Al deoxidation process and the smelting route is Electric furnace (EAF) → Ladle refining furnace (LF) → Vacuum refining furnace (VD) → Continuous casting (CC), the source of [Mg] in molten steel during smelting, the transformation process of inclusions from MgO-Al_2O_3 to liquid CaO-MgO-Al_2O_3, and the removal mechanism of inclusions have never been specifically studied.

In the present work, the samples were taken at ladle furnace (LF) and vacuum degassing (VD) in a plant trial to reveal the source of [Mg] in liquid steel and the formation mechanism of CaO-MgO-Al_2O_3 inclusions. The removal mechanism of low-melting-point CaO-MgO-Al_2O_3 inclusions was studied by kinetic calculation. The morphology, composition, and numbers of inclusions in samples were determined by scanning electron microscopy and energy dispersive spectroscopy.

2. Methodology

2.1. Experimental Procedures

The industrial trials of 42CrMo4 steel were carried out in a steel plant. The steelmaking route is Electric furnace (EAF) → Ladle refining furnace (LF) → Vacuum refining furnace (VD) → Continuous casting (CC), as shown in Figure 1. Al blocks are used as deoxidizers for strong deoxidation after steel tapping in the electric furnace, Si-Mn alloy and ferrochrome alloy are used for alloying, and then lime is added for slagging. The slag used in the LF stage has high basicity and strong reducing properties, and its binary basicity can reach around 7. In vacuum treatment in the VD stage at vacuum degrees below 66.7 pa, argon gas is passed through for stirring to promote rapid removal of hydrogen from the steel under vacuum conditions. At the end of the vacuum, to make the steel composition qualified, Al wire is added to increase the Al content in the molten steel to about 0.021%. The argon blowing pressure and flow rate are adjusted for a soft blowing operation, which lasts a total of 20 min. Finally, the liquid steel is transported to the continuous casting platform for casting. The entire sampling process was carried out for a certain furnace: LF refining in the early stage, LF refining in the middle stage, LF refining at the end, VD vacuum at the end (sampling immediately after VD vacuum), soft blowing for 10 min, and soft blowing at the end (for 20 min), respectively marked as samples 1~6. The slag was sampled at the beginning of LF refining, the middle of LF refining, the end of LF refining and the end of VD vacuum, respectively marked as samples 1~4. Since the composition of the slag after VD vacuum was basically unchanged, the slag at the soft blowing stage was not sampled.

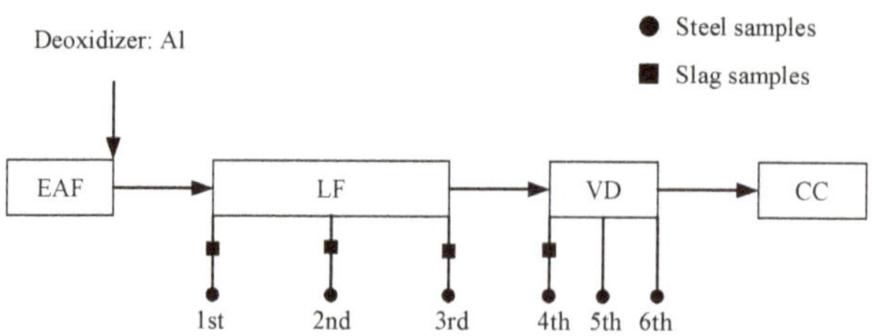

Figure 1. Schematic illustration of smelting process and sampling locations.

2.2. Composition Analysis and Inclusion Observation

The steel sample is processed into a round bar (Φ 5 × 50 mm), and the total oxygen content (T.O) in the sample is detected by the inert gas fusion-infrared absorptiometry method. The contents of Ca and Mg in steel were determined by the inductively coupled plasma optical emission spectrometry method (ICP-OES, Thermo Fisher, Waltham, MA, USA) and other alloy elements using the direct reading spectrometer ARL8860 (Thermo Fisher, Waltham, MA, USA) for testing; the main components of the sample were as shown in Table 1. The chemical composition of the slag was detected by X-ray fluorescence analysis (XRF, PANalytical B.V., Almelo, The Netherlands); the results were as shown in Table 2, which shows the slag basicity was around 7. In order to facilitate observation and statistics, the obtained steel samples were cut to obtain metallographic samples of 10 mm × 10 mm × 10 mm. After grinding and polishing, scanning electron microscopy (FEI, Hillsboro, OR, USA) and energy dispersive spectrometry (FEI, Hillsboro, OR, USA) were used for scanning statistics under fixed magnification, and the types and chemical components of inclusions obtained were analyzed.

Table 1. Chemical compositions of 42CrMo4 steel (mass percent).

Sample No.	C	Si	Mn	S	Cr	Mo	Al	Ca	Mg	T.O
1	0.280	0.158	0.618	0.0040	0.879	0.132	0.036	0.0015	0.0006	0.0024
2	0.403	0.204	0.678	0.0022	0.985	0.156	0.024	0.0009	0.0006	0.0029
3	0.408	0.213	0.677	0.0019	0.982	0.155	0.019	0.0014	0.0007	0.0036
4	0.418	0.208	0.670	0.0018	0.986	0.154	0.008	0.0017	0.0006	0.0027
5	0.415	0.209	0.675	0.0017	0.986	0.156	0.021	0.0019	0.0005	0.0023
6	0.404	0.208	0.672	0.0017	0.978	0.155	0.021	0.0010	0.0005	0.0014

Table 2. Chemical compositions of slag (mass percent).

Sample No.	CaO	SiO_2	Al_2O_3	MgO
1	58.6	8.57	24.2	5.33
2	56.4	7.92	26.5	5.87
3	55.4	8.26	27.2	5.93
4	54.2	8.7	28.2	6.31

3. Results

3.1. Characterization of Inclusions

By scanning the surface of each sample, it was found that the main inclusions in the steel were MgO-Al_2O_3 inclusions with irregular shapes in the early stage of LF smelting, as shown in Figure 2a. The mass fraction of each component of the inclusions is marked in the figure, and the mass fraction of MgO was lower than that of Al_2O_3. The elemental mapping of typical MgO-Al_2O_3 inclusions is shown in Figure 3a. Mg, Al, and O elements were evenly distributed in the inclusions. In the middle stage of LF smelting, the main types of inclusions remained unchanged and were still MgO-Al_2O_3 inclusions, but CaO began to appear in some inclusions, as shown in Figure 2(b2). The elemental mapping of typical inclusions is shown in Figure 3b. The distribution of Mg and Al was relatively uniform, whereas Ca was unevenly distributed in the outer layer of inclusion. In this smelting stage, a small number of spherical inclusions with high CaO content also began to appear in the molten steel, as shown in Figure 2(b3). At the end of LF smelting, CaO generally appeared in inclusions, and the shape of CaO-MgO-Al_2O_3 inclusions containing a small amount of CaO was irregular, as shown in Figure 2(c1). With the further increase in CaO content in inclusions, the morphology of inclusions began to change from irregular shape to spherical inclusions with smooth external contour, as shown in Figure 2(c2). The elemental mapping of typical inclusions is shown in Figure 3c. In the outer layer of the inclusions, the Ca element further accumulated, whereas the Mg element disappeared, and the distribution of the Al element decreased slightly. In the center of inclusion, the distribution of Mg and

Al was more concentrated, but the Ca element was not present. At the end of VD vacuum refining, the main type of inclusions in molten steel was CaO-MgO-Al_2O_3 inclusions with a spherical shape, and the content of MgO in the inclusions decreased obviously, as shown in Figure 2d. The elemental mapping of typical CaO-MgO-Al_2O_3 inclusions is shown in Figure 3d. It can be seen that the distribution of Ca and Al elements in the inclusion is uniform. After 10 and 20 min of VD soft blowing, the main types of inclusions were the same as those at the end of the VD vacuum. The main inclusions were CaO-MgO-Al_2O_3 inclusions containing a small amount of MgO, as shown in Figure 2. The elemental mapping of typical CaO-MgO-Al_2O_3 inclusions is shown in Figure 3e,f. It can be seen that the distribution of Ca and Al elements in the inclusions was uniform, but the distribution of Mg elements was uniform or a with small amount of aggregation.

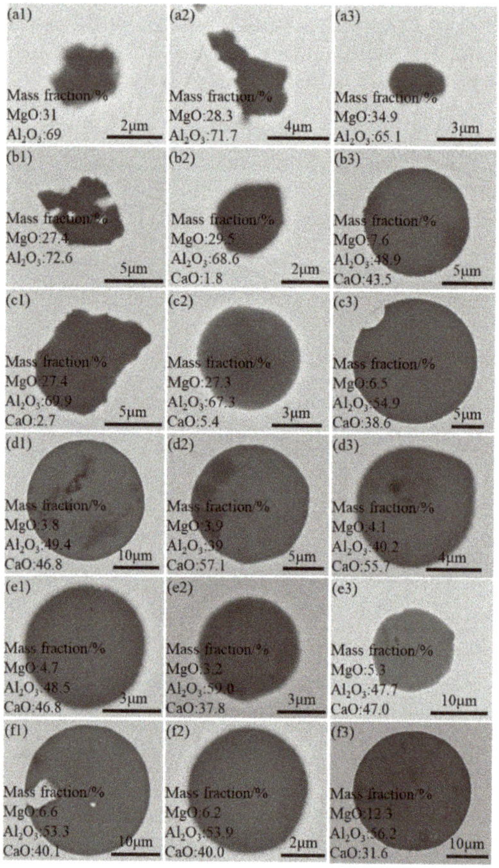

Figure 2. Morphology and composition of typical inclusions encountered in samples: (**a1**) a MgO-Al_2O_3 inclusion in sample 1; (**a2**) a MgO-Al_2O_3 inclusion in sample 1; (**a3**) a MgO-Al_2O_3 inclusion in sample 1; (**b1**) a MgO-Al_2O_3 inclusion in sample 2; (**b2**) a CaO-MgO-Al_2O_3 inclusion in sample 2; (**b3**) a CaO-MgO-Al_2O_3 inclusion in sample 2; (**c1**) a CaO-MgO-Al_2O_3 inclusion in sample 3; (**c2**) a CaO-MgO-Al_2O_3 inclusion in sample 3; (**c3**) a CaO-MgO-Al_2O_3 inclusion in sample 3; (**d1**) a CaO-MgO-Al_2O_3 inclusion in sample 4; (**d2**) a CaO-MgO-Al_2O_3 inclusion in sample 4; (**d3**) a CaO-MgO-Al_2O_3 inclusion in sample 4; (**e1**) a CaO-MgO-Al_2O_3 inclusion in sample 5; (**e2**) a CaO-MgO-Al_2O_3 inclusion in sample 5; (**e3**) a CaO-MgO-Al_2O_3 inclusion in sample 5; (**f1**) a CaO-MgO-Al_2O_3 inclusion in sample 6; (**f2**) a CaO-MgO-Al_2O_3 inclusion in sample 6; (**f3**) a CaO-MgO-Al_2O_3 inclusion in sample 6.

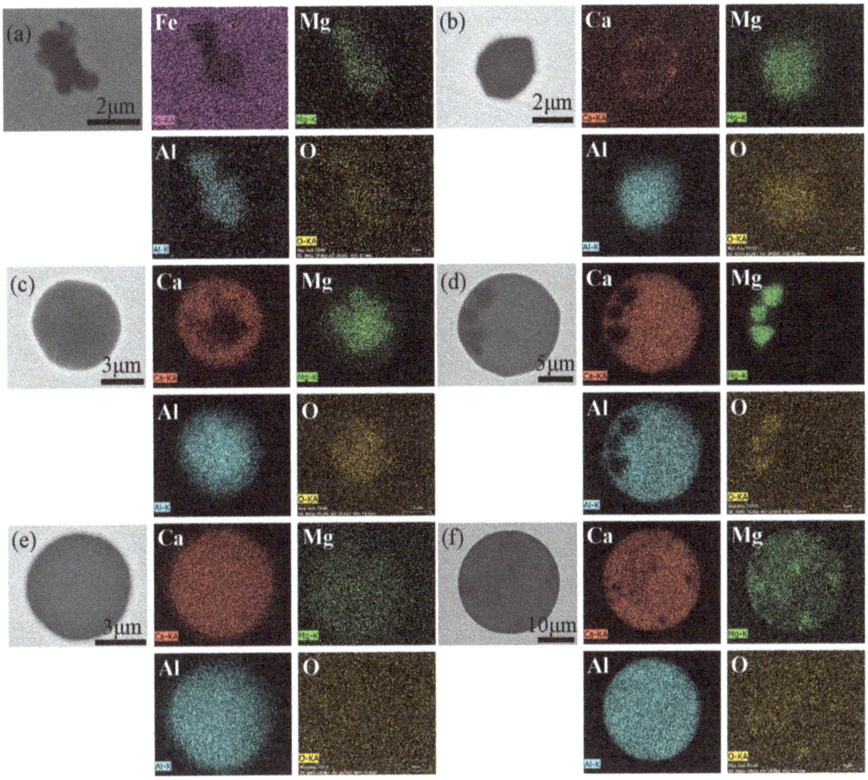

Figure 3. Elemental mapping of typical inclusions in samples: (**a**) a typical MgO-Al_2O_3 inclusion in sample 1; (**b**) a typical MgO-Al_2O_3 inclusion in sample 2; (**c**) a typical CaO-MgO-Al_2O_3 inclusion in sample 3; (**d**) a typical CaO-MgO-Al_2O_3 inclusion in sample 4; (**e**) a typical CaO-MgO-Al_2O_3 inclusion in sample 5; (**f**) a typical CaO-MgO-Al_2O_3 inclusion in sample 6.

3.2. Composition Distribution of Inclusions

The inclusion compositions observed at each stage were marked in the CaO-MgO-Al_2O_3 ternary phase diagram, as shown in Figure 4. The ternary phase diagram of CaO-MgO-Al_2O_3 at 1600 °C was calculated by FactSage8.0 thermodynamics software (version 8.0, CRCT, Montreal, Canada), and the red line area denotes the pure liquid oxide area. Aluminum blocks were used for deoxidation in the smelting process, and Al_2O_3 inclusions would be rapidly formed in the molten steel. However, at the initial stage of LF refining, the main type of inclusions in liquid the steel was MgO-Al_2O_3 inclusions. With the development of LF refining, a new component, CaO, began to appear in the MgO-Al_2O_3 inclusions. Many studies have found that the impure alloys [32] and the reaction between slag and steel [33,34] transfer [Ca] to the steel, resulting in the appearance of CaO elements in the inclusions of the non-calcium-treated LF stage. As can be seen from Figure 4, the composition of the inclusions approached the liquid oxide region as the smelting proceeded, and the morphology of the inclusions also changed from irregular to a spherical shape. After the end of the VD vacuum treatment, MgO-Al_2O_3 inclusions in liquid steel were mostly transformed into spherical liquid low-melting-point CaO-MgO-Al_2O_3 inclusions, which were principally in the pure liquid oxide region. After soft blowing, the main type of inclusions remained unchanged.

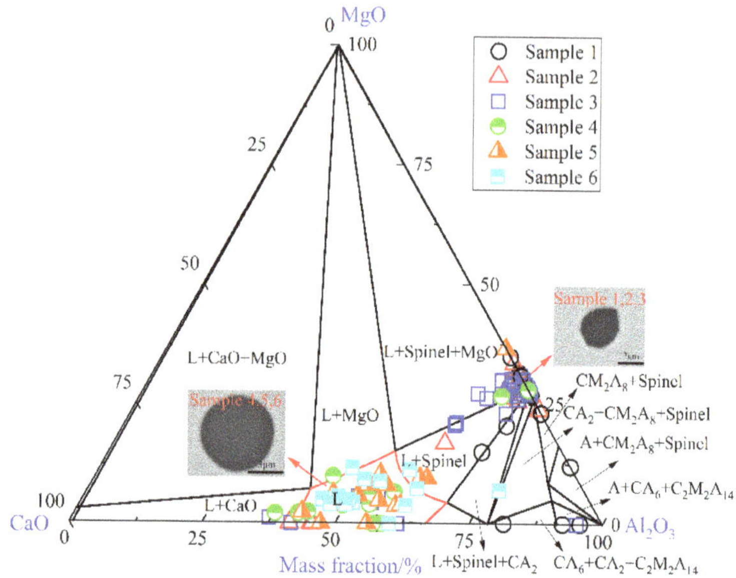

Figure 4. Composition distributions (mass fraction) of inclusions in CaO-MgO-Al_2O_3 phase diagrams. (L: Liquid Oxide; CA_2: CaO-$2Al_2O_3$; CM_2A_8: CaO-$2MgO$-$8Al_2O_3$; CM_2A_{14}: CaO-$2MgO$-$14Al_2O_3$; CA_6: CaO-$6Al_2O_3$; Spinel:MgO-Al_2O_3).

3.3. *Quantity Distribution of Inclusions*

It can be seen from Figure 4 that the inclusions are mainly divided into three types: MgO-Al_2O_3 inclusions without CaO, CaO-MgO-Al_2O_3 inclusions in the solid–liquid two-phase region, and CaO-MgO-Al_2O_3 inclusions in the pure liquid phase region. The content of CaO in the CaO-MgO-Al_2O_3 inclusions in the solid–liquid two-phase region ranged from 0 to 30%. The number of inclusions of different types in the samples was further analyzed, as shown in Figure 5. In this paper, all inclusions larger than 3 μm on the surface of each sample were counted, so the total number of inclusions in each sample was different. In samples 1~3, the main type of inclusions was MgO-Al_2O_3 inclusions, but four CaO-containing inclusions were found in sample 1, and the CaO content was less than 30%. The number of inclusions in sample 2 increased to 10 and included two liquid CaO-MgO-Al_2O_3 inclusions. In sample 3, the number of liquid CaO-MgO-Al_2O_3 inclusions increased to 7, whereas the number of MgO-Al_2O_3 inclusions decreased slightly. In samples 4–6, the main inclusions changed into liquid CaO-MgO-Al_2O_3 inclusions. In sample 4, the number of MgO-Al_2O_3 inclusions decreased to 1. The number of CaO-MgO-Al_2O_3 inclusions containing less than 30% CaO also decreased significantly. The number of liquid CaO-MgO-Al_2O_3 inclusions increased significantly to 17. As for samples 5 and 6, the number of liquid CaO-MgO-Al_2O_3 inclusions was basically unchanged, at 17 and 15, respectively, whereas the number of other types of inclusions was very small, and no MgO-Al_2O_3 inclusions were found in sample 6.

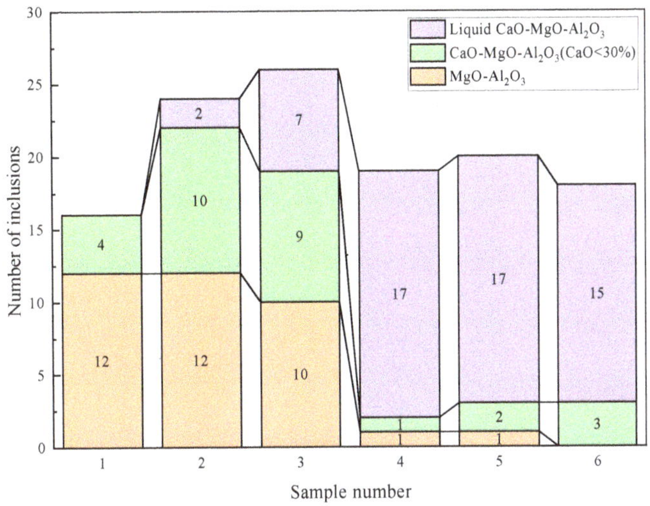

Figure 5. Number of inclusions of different types in all samples.

4. Discussion

4.1. Thermodynamic Analysis of [Mg] Sources in Molten Steel

After the reaction of [Al] with refractory and slag in steel, the refractory and slag will transfer [Mg] to the steel, which is the fundamental cause of the formation of $MgO\text{-}Al_2O_3$ inclusions in this experiment. During the smelting process, the reactions between liquid steel and refractory, liquid steel, and slag are as follows [10]:

$$3MgO(s)_{\text{in refractory}} + 2[Al] = (Al_2O_3)_{\text{in slag}} + 3[Mg] \tag{1}$$

$$3(MgO)_{\text{in slag}} + 2[Al] = (Al_2O_3)_{\text{in slag}} + 3[Mg] \tag{2}$$

The reaction between steel and slag and refractory erosion in 42CrMo4 steel was theoretically calculated by using FactSage8.0 thermodynamic software. The actual composition of liquid steel detected in Table 1 was adopted. In order to simplify the calculation, the following model assumptions were used: (1) the reaction between molten steel and slag reached equilibrium; (2) the reaction between liquid steel and refractory was balanced; (3) the temperature was constant at 1600 °C; (4) the erosion of refractory by slag was not considered. The mass fraction ratio of molten steel to slag and molten steel to refractory was 100:1 during calculation. The calculation results are shown in Figure 6. To increase the number of samples and thus improve the accuracy of the model, we conducted five experiments under the same conditions, and full-process sampling was performed for each experiment. A~E are the five experiments conducted, and 1~4 are the sampling numbers for LF refining in the early stage, LF refining in the middle stage, LF refining at the end, and VD vacuum at the end for each experiment. It can be seen that both slag and refractory transferred [Mg] to liquid steel. In the LF refining process, the transfer of [Mg] from refractory to molten steel was more than slag, which played a major role. With the progress of smelting, the content of [Mg] transferred by refractory gradually decreased, which was related to the decrease in [Al] content in molten steel. The decrease in [Al] content reduced the corrosion ability of liquid steel to refractory. The transfer of [Mg] content from slag to molten steel gradually increased to a stable level. In the VD refining process, molten steel and slag were fully stirred. Slag-transferred [Mg] content to molten steel was higher than that transferred by refractory, which played a major role.

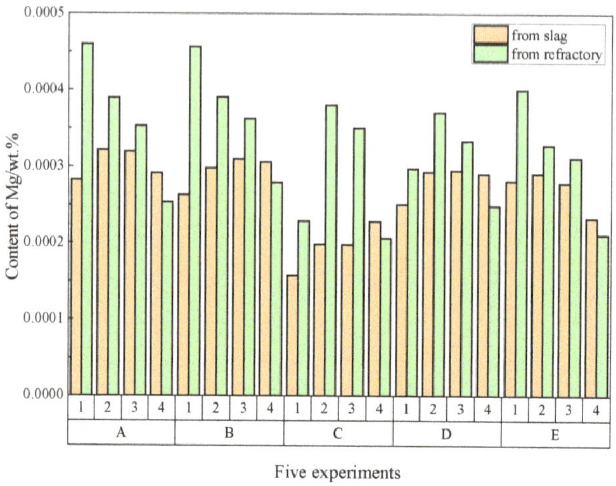

Figure 6. The content of magnesium provided by slag and refractory to molten steel at 1600 °C.

In order to verify the accuracy of the thermodynamic calculation, error analysis was conducted on the calculation results of the five furnace experiments, as shown in Figure 7. Mg is a trace element, and Kang [35] pointed out that there is 0.0001–0.0002% error in the determination of Mg content by the ICP method. The result of thermodynamic calculation is compared with the result of experimental detection, and the fluctuation of the value is not more than 0.00015%. Therefore, the thermodynamic calculation results can reflect the content of [Mg] in molten steel well.

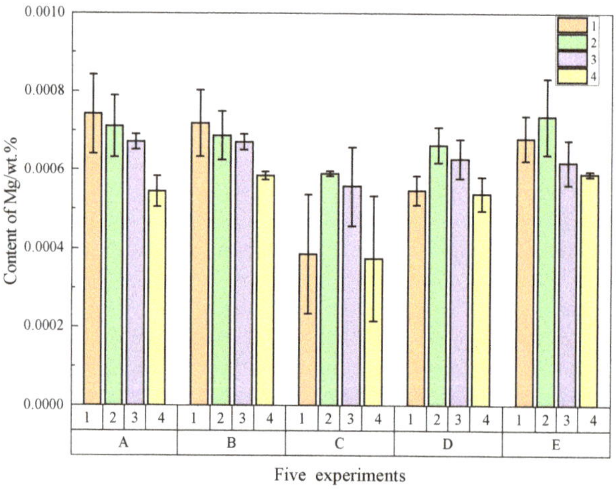

Figure 7. Error analysis of calculation results of five furnaces.

4.2. Formation Mechanism of CaO-MgO-Al_2O_3 Inclusions

After slag and refractories transferred [Mg] into liquid steel, MgO-Al_2O_3 inclusions were rapidly generated in liquid steel, so a large number of such inclusions were observed in the early stage of LF refining. The Mg-Al-O equilibrium phase diagram of 42CrMo4 steel at 1600 °C was calculated by FactSage8.0 software, as shown in Figure 8. The dotted line shows the dissolved oxygen contents in molten steel. Figure 9a contains four areas: the upper part is the MgO inclusion generation area, the middle part is the MgO-Al_2O_3 inclusion generation area, the lower part is the Al_2O_3 inclusion generation area, and the lower-left corner is the liquid steel area. It can be seen that the formation area of MgO-Al_2O_3 inclusions is the largest. A small amount of [Mg] contents in liquid steel could form MgO-Al_2O_3 inclusions. Figure 8b,c shows the effect of adding 1 ppm and 3 ppm [Ca] to liquid steel on inclusions, respectively. There are four areas: the upper part is the MgO inclusion generation area, the middle part is the MgO-Al_2O_3 inclusion generation area, the lower part is the CaO-MgO-Al_2O_3 liquid oxide generation area, and the lower-left corner is the liquid steel area.

The composition of liquid steel in LF and VD is marked in Figure 8. Without [Ca], as shown in Figure 8a, the composition points are all located in the formation area of MgO-Al_2O_3 inclusions, indicating that a large number of MgO-Al_2O_3 inclusions were formed in the molten steel under this smelting condition. When 1 ppm [Ca] existed in the liquid steel, as shown in Figure 8b, the composition points were still all located in the area of MgO-Al_2O_3 inclusions. However, the Al_2O_3 inclusion formation region disappeared. The CaO-MgO-Al_2O_3 liquid oxide formation region appeared. The MgO-Al_2O_3 inclusion region became smaller. It can be found that 1 ppm [Ca] was not enough to convert inclusions in 42CrMo4 steel to CaO-MgO-Al_2O_3 liquid oxide. When 3 ppm [Ca] existed in the liquid steel, as shown in Figure 8c, the MgO-Al_2O_3 inclusion formation area in the phase diagram almost disappeared and only occurred when the dissolved oxygen in the liquid steel was high. The formation area of CaO-MgO-Al_2O_3 liquid oxide in the phase diagram was further expanded. The composition points of liquid steel are all located in this region, indicating that CaO-MgO-Al_2O_3 liquid oxide was stably generated in 42CrMo4 steel when there was 3 ppm [Ca] in liquid steel. It can be found that increasing different calcium content in molten steel had no effect on the MgO inclusion formation region, but increasing only 1 ppm calcium significantly affected the MgO-Al_2O_3 inclusion and Al_2O_3 inclusion formation area. Thus, a slight increase in calcium could change the type of inclusions in molten steel.

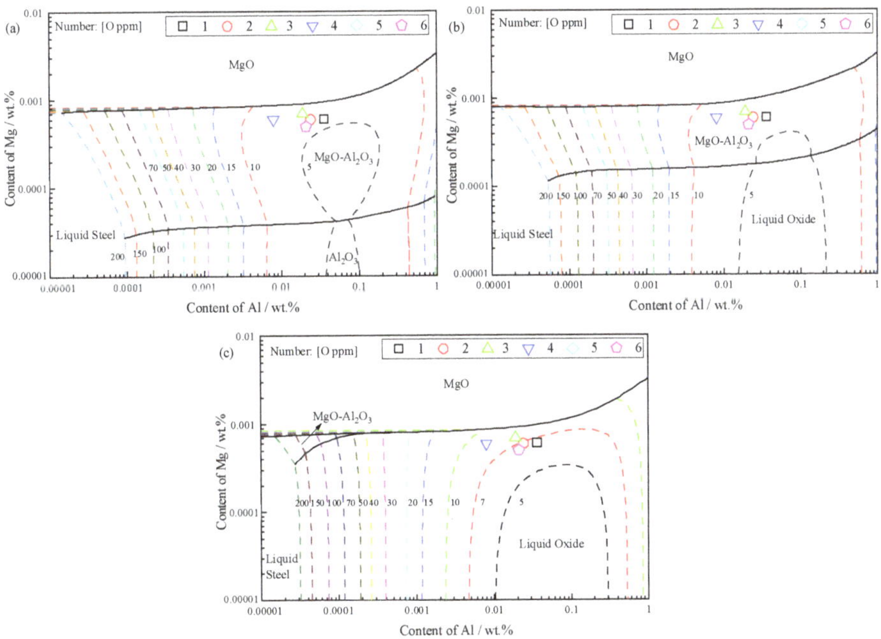

Figure 8. Effect of different calcium content on MgO-Al_2O_3 inclusions: (**a**) no calcium; (**b**) 1 ppm [Ca]; (**c**) 3 ppm [Ca] (C: 0.4%-Cr: 0.98%-Mo: 0.16%).

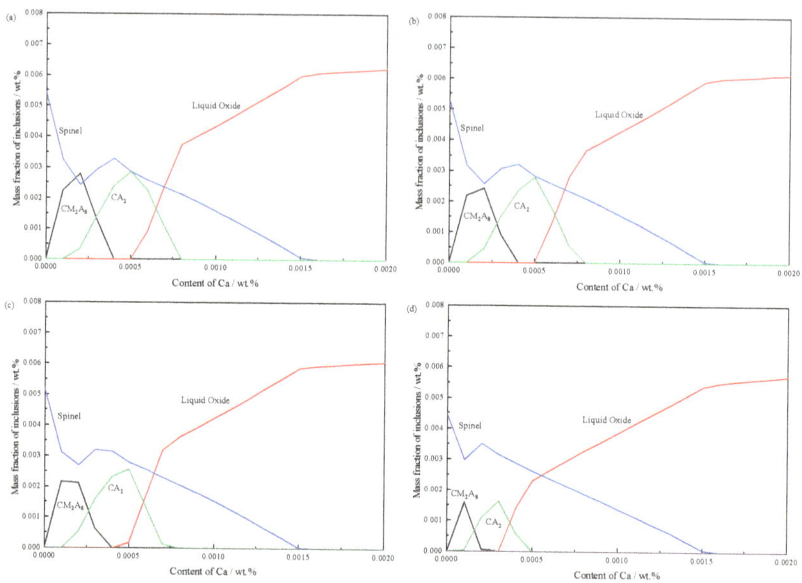

Figure 9. Effect of calcium content on inclusions in smelting process at 1600 °C: (**a**) Al: 0.036%; (**b**) Al: 0.024%; (**c**) Al: 0.02%; (**d**) Al: 0.008%; C: 0.4%; Cr: 0.98%; Mo: 0.16%; Mg: 0.0006%; O: 0.003%; Spinel: $MgO\text{-}Al_2O_3$; CM_2A_8: $CaO\text{-}2MgO\text{-}8Al_2O_3$; CA_2: $CaO\text{-}2Al_2O_3$; C_2S: $CaO\text{-}2SiO_2$.

According to the observed inclusions distribution, $MgO\text{-}Al_2O_3$ inclusions gradually changed to $CaO\text{-}MgO\text{-}Al_2O_3$ inclusions from LF to VD. The CaO content in the inclusions increased gradually until the end of VD soft blowing, and most of the inclusions were in the liquid oxide region. The calcium and aluminum in the liquid steel fluctuated greatly according to the composition of the liquid steel detected. Therefore, in order to explore the effect of calcium content on inclusion transformation in liquid steel, FactSage8.0 software was used to calculate the effect of calcium content on inclusions in liquid steel under different aluminum content conditions, as shown in Figure 9. The calculation range of calcium content was 0~0.004%; Mg content was 0.0006%; the total oxygen content was 0.003%. It can be seen from Figure 9a that with the increase in calcium content, the content of $MgO\text{-}Al_2O_3$ inclusions in molten steel gradually decreased, and $CaO\text{-}2MgO\text{-}8Al_2O_3$ and $CaO\text{-}2Al_2O_3$ inclusions began to appear. The content of $CaO\text{-}2MgO\text{-}8Al_2O_3$ inclusions began to decrease when the calcium content increased to 2 ppm, whereas the content of $CaO\text{-}2Al_2O_3$ inclusions continued to increase. When the calcium content increased to 5 ppm, the content of $CaO\text{-}2Al_2O_3$ inclusions began to decrease and liquid oxides began to appear. When the calcium content reached 16ppm, the $MgO\text{-}Al_2O_3$ inclusions disappeared completely and only liquid low-melting-point oxides remained in the molten steel. The transition route of inclusions was: $MgO\text{-}Al_2O_3 \rightarrow CaO\text{-}2MgO\text{-}8Al_2O_3 \rightarrow CaO\text{-}2MgO\text{-}8Al_2O_3 + CaO\text{-}2Al_2O_3 \rightarrow$ Liquid $CaO\text{-}MgO\text{-}Al_2O_3$. The content of CaO in inclusions increased gradually. The inclusions in Figure 9b–d had the same transition route as Figure 9a, except that the quality of inclusions generated under different aluminum contents was different.

4.3. Removal Mechanism of Inclusions

According to the observation and analysis of inclusions in 42CrMo4 steel, it was shown that low-melting-point $CaO\text{-}MgO\text{-}Al_2O_3$ inclusions became the main inclusions in liquid steel after LF, VD, and soft blowing. Soft blowing treatment could not effectively promote the removal of such inclusions. The solid $MgO\text{-}Al_2O_3$ inclusions disappeared after refining, indicating that the removal effects of different types of inclusions in molten steel were different. The removal process of inclusions was divided into three steps: polymerization growth, floating, and separation at the interface between steel and slag. The third step determines whether the inclusion could be removed from the liquid steel [26]. This section discusses the displacement, velocity, and force variation of different inclusions at the interface between steel and slag.

Based on the Strandh model [28,29], the force analysis diagram of inclusions at the interface between steel and slag is shown in Figure 10. Inclusion in the interface between steel and slag sported by the four forces worked together: the buoyancy force F_b, the drag force F_d, the added mass force F_f, and the rebound force F_r. Inclusions of movement followed Newton's second law:

$$F_a = ma = F_b + F_r + F_d + F_f = \frac{4}{3}\pi {R_I}^3 \rho_I \frac{d^2 Z}{dt^2} \quad (3)$$

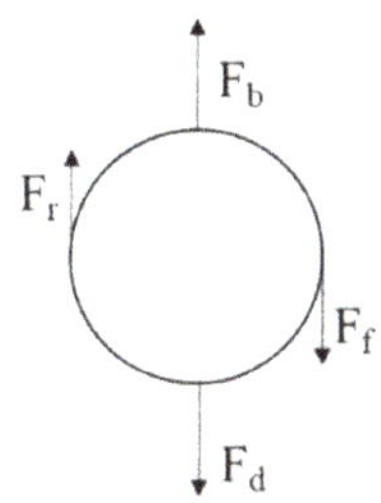

Figure 10. The force analysis diagram of inclusions.

The size of the buoyancy force, the drag force, the added mass force, and the rebound force could be calculated according to Equations (4)–(7):

$$F_r = 2\pi R \sigma_{MS}\left(\frac{Z}{R_I} - 1 - \frac{\sigma_{IM} - \sigma_{IS}}{\sigma_{MS}}\right) \quad (4)$$

$$F_d = 4\pi {R_I}^2 \mu_S A \cdot \left(\left(\frac{\mu_M}{\mu_S} - 1\right)\frac{Z^2}{{R_I}^2} - 2\left(\frac{\mu_M}{\mu_S} - 1\right)\frac{Z}{R_I} + \frac{\mu_M}{\mu_S}\right)\frac{dZ}{dt} \quad (5)$$

$$F_f = \frac{2}{3}\pi R_I^3 \rho_S \left(\frac{1}{4}\left(\frac{\rho_M}{\rho_S} - 1\right)\frac{Z^3}{{R_I}^3} - \frac{3}{4}\left(\frac{\rho_M}{\rho_S} - 1\right)\frac{Z^2}{{R_I}^2} + \frac{\rho_M}{\rho_S}\right)\frac{d^2 Z}{dt^2} \quad (6)$$

$$F_b = \frac{4}{3}\pi R_I^3 g\left(\left(\frac{1}{4}\left(\frac{\rho_M}{\rho_S} - 1\right)\frac{Z^3}{{R_I}^3} - \frac{3}{4}\left(\frac{\rho_M}{\rho_S} - 1\right)\frac{Z^2}{{R_I}^2} + \frac{\rho_M}{\rho_S}\right)\rho_S - \rho_I\right) \quad (7)$$

where R is the radius of inclusion, σ_{MS} is the interfacial tension between molten steel and slag, σ_{IM} is the interfacial tension between inclusions and liquid steel, σ_{IS} is the interfacial tension between inclusions and slag, μ_M is the viscosity of liquid steel, μ_S is the viscosity of slag, ρ_M is the density of liquid steel, ρ_S is the slag density, ρ_I is the density of inclusions, Z is the displacement of inclusions at the interface between steel and slag, and t is time. In this model, it was assumed that the initial position of the inclusions was below the interface. When the displacement value reached twice the radius of the inclusions, it was considered that the inclusions broke away from the interface and completely entered the slag.

According to Newton's second law, a second-order differential equation could be obtained by simultaneous operation of Equations (3)–(6), as shown in Equation (8). The change in inclusion displacement with time could be obtained by solving the equation.

$$\frac{d^2Z}{dt^2} = \frac{2\left(\rho_S\left(\frac{1}{4}\left(\frac{\rho_M}{\rho_S}-1\right)\frac{Z^3}{R_I{}^3}-\frac{3}{4}\left(\frac{\rho_M}{\rho_S}-1\right)\frac{Z^2}{R_I{}^2}+\frac{\rho_M}{\rho_S}\right)-\rho_I\right)g}{\left(\rho_S\left(\frac{1}{4}\left(\frac{\rho_M}{\rho_S}-1\right)\frac{Z^3}{R_I{}^3}-\frac{3}{4}\left(\frac{\rho_M}{\rho_S}-1\right)\frac{Z^2}{R_I{}^2}+\frac{\rho_M}{\rho_S}\right)+2\rho_I\right)} - \frac{3\sigma_{MS}\left(\frac{Z}{R_I}-1-\frac{\sigma_{IM}-\sigma_{IS}}{\sigma_{MS}}\right)}{R_I^2\left(\rho_S\left(\frac{1}{4}\left(\frac{\rho_M}{\rho_S}-1\right)\frac{Z^3}{R_I{}^3}-\frac{3}{4}\left(\frac{\rho_M}{\rho_S}-1\right)\frac{Z^2}{R_I{}^2}+\frac{\rho_M}{\rho_S}\right)+2\rho_I\right)} - 6A\mu_S\sqrt{R_I^2\left(\rho_S\left(\frac{1}{4}\left(\frac{\rho_M}{\rho_S}-1\right)\frac{Z^3}{R_I{}^3}-\frac{3}{4}\left(\frac{\rho_M}{\rho_S}-1\right)\frac{Z^2}{R_I{}^2}+\frac{\rho_M}{\rho_S}\right)+2\rho_I\right)} \times\left(\left(\frac{\mu_M}{\mu_S}-1\right)\frac{Z^2}{R_I{}^2}-2\left(\frac{\mu_M}{\mu_S}-1\right)\frac{Z}{R_I}+\frac{\mu_M}{\mu_S}\right)\frac{dZ}{dt} \quad (8)$$

The low-melting-point CaO-MgO-Al_2O_3 inclusions detected in the smelting process were substituted into Equation (7) for solution, as shown in Figure 11. It can be seen that the displacement of the low-melting-point CaO-MgO-Al_2O_3 inclusions generated in the smelting process at the interface between steel and slag was less than 2R. This kind of inclusion could not be removed from the molten steel, but oscillated at the interface between steel and slag, and finally stayed there. In actual production, the inclusions were likely to return to the molten steel due to the fluctuation of the interface.

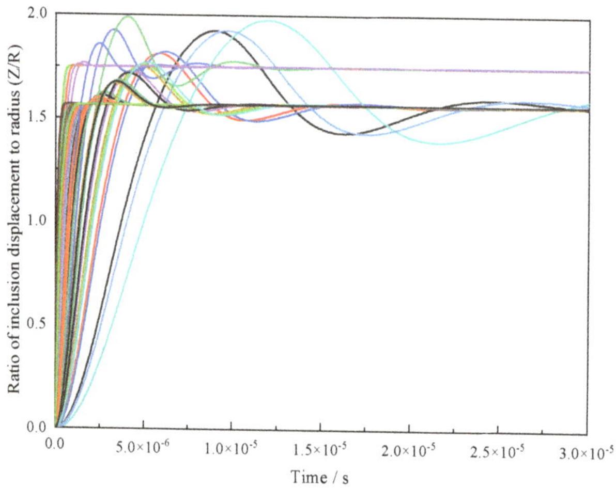

Figure 11. Displacement variation of low-melting-point CaO-MgO-Al_2O_3 inclusions at the interface between steel and slag.

Since a large number of MgO-Al_2O_3 inclusions were observed in the LF refining process, the removal process of such inclusions at the interface was calculated, as shown in Figure 12. The displacement changes of MgO-Al_2O_3 inclusion and low-melting-point CaO-MgO-Al_2O_3 inclusion at the interface were compared under the same radius (15 μm), molten steel and slag conditions. The values of interfacial tension, density, and viscosity data required for the calculations were calculated according to the study by Xuan et al. [27], as shown in Table 3. The displacement of the MgO-Al_2O_3 inclusion reached 2R at 5.04×10^{-6} s, whereupon it could completely separate from the molten steel and entered the slag. The displacement of the low-melting-point CaO-MgO-Al_2O_3 inclusion was obviously smaller than that of the MgO-Al_2O_3 inclusion at the same time, and reached the maximum displacement at 1.2×10^{-5} s. However, at this time, the inclusion could not completely break away from the molten steel. Then, the displacement decreased, and the inclusion oscillated at the interface until it stayed at the interface. Finally, the displacement of the inclusion was 1.57 R. The MgO-Al_2O_3 inclusion separated from the liquid steel and entered the slag phase faster than the low-melting-point CaO-MgO-Al_2O_3 inclusions. Therefore, the low-melting-point

CaO-MgO-Al_2O_3 inclusion was not easily removed from liquid steel, which was also the reason why a large number of such inclusions were still seen after VD.

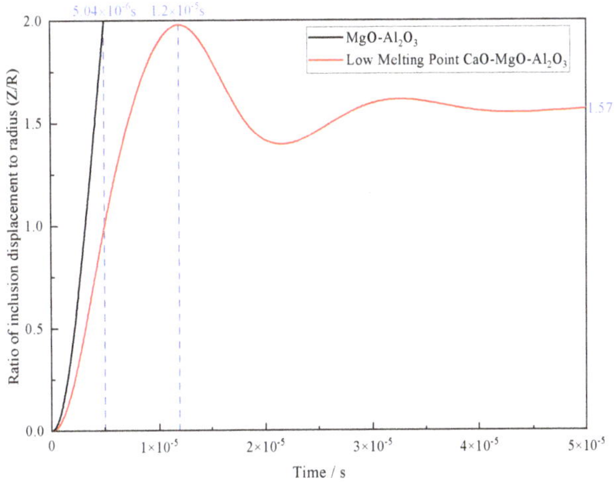

Figure 12. The displacement of different inclusions at the interface between steel and slag.

Table 3. Calculated values of inclusion radius, interfacial tension, density, and viscosity.

Inclusion Type	R_I (μm)	σ_{MS} (N/m)	σ_{IM} (N/m)	σ_{IS} (N/m)	μ_M (Pa•s)	μ_S (Pa•s)	ρ_M (kg/m^3)	ρ_S (kg/m^3)	ρ_I (kg/m^3)
Liquid CaO-MgO-Al_2O_3	15	1.36	1.34	0.57	0.0049	0.089	7001.65	2809.42	3055.52
Solid MgO-Al_2O_3	15	1.36	3.13	0.82	0.0049	0.089	7001.65	2809.42	3550 [36]

The velocity and resultant force changes of the low-melting-point CaO-MgO-Al_2O_3 inclusion in Figure 12 were further calculated and analyzed, as shown in Figure 13. It can be seen that the inclusion velocity decreased to 0 at 1.2×10^{-5} s, when the inclusion had the maximum displacement. As the resultant force on the inclusion was the resistance to inclusion removal, the value is -4×10^{-5} N. Under the action of the resultant force, the inclusion began to change from moving out of liquid steel to moving into liquid steel. Then, it oscillated until the velocity and resultant force changed to zero and stayed at the interface.

Under the same conditions of steel and slag, the low-melting-point CaO-MgO-Al_2O_3 inclusions and MgO-Al_2O_3 inclusions exhibited completely different movement behaviors at the interface of steel and slag, which was related to their interfacial tension [26]. Therefore, the interfacial tension of the two inclusions was calculated, as shown in Figure 14. The low-melting-point CaO-MgO-Al_2O_3 inclusions in the liquid oxide region had similar physical properties, so the mean value was used to represent them. It can be seen that the interfacial tension between low-melting-point CaO-MgO-Al_2O_3 inclusions and slag was slightly smaller than that between MgO-Al_2O_3 inclusions and slag under the same conditions. However, the interfacial tension between low-melting-point CaO-MgO-Al_2O_3 inclusions and molten steel was much less than that between MgO-Al_2O_3 inclusions and molten steel, which was less than 50% of the former. The large difference in the interfacial tension of the two inclusions resulted in different movement behaviors at the interface between steel and slag, which was the reason why the low-melting-point inclusions could not be removed from molten steel and became the main type of inclusions in molten steel.

Institutional Review Board Statement: Not applicable.

Informed Consent Statement: Not applicable.

Data Availability Statement: Not applicable.

Acknowledgments: The authors would like to express appreciation to the National Natural Science Foundation of China for the financial support, the State Key Laboratory of Advanced Metallurgy at the University of Science and Technology Beijing, and to Jiangsu Yonggang Group Co., Ltd., for academic and experimental assistance.

Conflicts of Interest: The authors declare no conflict of interest.

Nomenclature

F_a	The combined force on the inclusions (N)
F_b	The buoyancy force (N)
F_d	The drag force (N)
F_f	The added mass force (N)
F_r	The rebound force (N)
R_I	The radius of inclusions (μm)
σ_{MS}	The interfacial tension between molten steel and slag (N/m)
σ_{IM}	The interfacial tension between inclusions and liquid steel (N/m)
σ_{IS}	The interfacial tension between inclusions and slag (N/m)
μ_M	The viscosity of liquid steel (Pa•s)
μ_S	The viscosity of slag (Pa•s)
ρ_M	The density of liquid steel (kg/m^3)
ρ_S	The slag density (kg/m^3)
ρ_I	The density of inclusions (kg/m^3)
Z	The displacement of inclusions at the interface between steel and slag (μm)
t	The time (s)

References

1. He, P.; Hong, R.; Wang, H.; Lu, C. Fatigue life analysis of slewing bearings in wind turbines. *Int. J. Fatigue* **2018**, *111*, 233–242. [CrossRef]
2. Seleznev, M.; Wong, K.Y.; Stoyan, D.; Weidner, A.; Biermann, H. Cluster detection of non-metallic inclusions in 42CrMo4 steel. *Steel Res. Int.* **2018**, *89*, 1800216. [CrossRef]
3. Zhang, H.; Zhu, M.; Ji, S.; Zhang, J.; Fan, H. Numerical simulation and experimental study on laser hardening process of the 42CrMo4 Steel. *Adv. Mech. Eng.* **2021**, *13*, 16878140211044648. [CrossRef]
4. Escalero, M.; Blasón, S.; Zabala, H.; Torca, I.; Urresti, I.; Muniz-Calvente, M.; Fernández-Canteli, A. Study of alternatives and experimental validation for predictions of hole-edge fatigue crack growth in 42CrMo4 steel. *Eng. Struct.* **2018**, *176*, 621–631. [CrossRef]
5. Cubillas, D.; Olave, M.; Llavori, I.; Ulacia, I.; Larrañaga, J.; Zurutuza, A.; Lopez, A. Numerical analysis of the wind turbine pitch bearing raceway tribo-contact due to cyclic loading under constant pitch angle. In *Proceedings of the 8th International Conference on Fracture, Fatigue and Wear*; Abdel Wahab, M., Ed.; Springer: Singapore, 2021; pp. 757–769.
6. Yang, Z.G.; Li, S.X.; Zhang, J.M.; Zhang, J.F.; Li, G.Y.; Li, Z.B.; Hui, W.J.; Weng, Y.Q. The fatigue behaviors of zero-inclusion and commercial 42CrMo steels in the super-long fatigue life regime. *Acta Mater.* **2004**, *52*, 5235–5241. [CrossRef]
7. Zhang, J.; Li, S.; Yang, Z.; Li, G.; Hui, W.; Weng, Y. Influence of inclusion size on fatigue behavior of high strength steels in the gigacycle fatigue regime. *Int. J. Fatigue* **2007**, *29*, 765–771. [CrossRef]
8. Tsunekage, N.; Hashimoto, K.; Fujimatsu, T.; Hiraoka, K.; Beswick, J.; Dean, S.W. Initiation behavior of crack originated from non-metallic inclusion in rolling contact fatigue. *J. ASTM Int.* **2010**, *7*, 97–110. [CrossRef]
9. Liu, C.; Gao, X.; Ueda, S.; Guo, M.; Kitamura, S. Composition changes of inclusions by reaction with slag and refractory: A review. *ISIJ Int.* **2020**, *60*, 1835–1848. [CrossRef]
10. Park, J.H.; Todoroki, H. Control of $MgO{\cdot}Al_2O_3$ spinel inclusions in stainless steels. *ISIJ Int.* **2010**, *50*, 1333–1346. [CrossRef]
11. Kumar, D.; Pistorius, P.C. A study on calcium transfer from slag to steel and its effect on modification of alumina and spinel inclusions. In *Advances in Molten Slags, Fluxes, and Salts: Proceedings of the 10th International Conference on Molten Slags, Fluxes and Salts 2016*; Reddy, R.G., Chaubal, P., Pistorius, P.C., Pal, U., Eds.; Springer International Publishing: Cham, Switzerland, 2016; pp. 145–153.
12. Yang, S.; Wang, Q.; Zhang, L.; Li, J.; Peaslee, K. Formation and modification of $MgO{\cdot}Al_2O_3$-based inclusions in alloy steels. *Metall. Mater. Trans. B* **2012**, *43*, 731–750. [CrossRef]
13. Yin, X.; Sun, Y.H.; Yang, Y.D.; Bai, X.F.; Deng, X.X.; Barati, M.; McLean, A. Inclusion Evolution during refining and continuous casting of 316L stainless steel. *Ironmak. Steelmak.* **2016**, *43*, 533–540. [CrossRef]

14. Park, J.H.; Kang, Y. Inclusions in stainless steels—A review. *Steel Res. Int.* **2017**, *88*, 1700130. [CrossRef]
15. Verma, N.; Pistorius, P.C.; Fruehan, R.J.; Potter, M.S.; Oltmann, H.G.; Pretorius, E.B. Calcium modification of spinel inclusions in aluminum-killed steel: Reaction steps. *Metall. Mater. Trans. B* **2012**, *43*, 830–840. [CrossRef]
16. Okuyama, G.; Yamaguchi, K.; Takeuchi, S.; Sorimachi, K. Effect of slag composition on the kinetics of formation of Al_2O_3-MgO inclusions in aluminum killed ferritic stainless steel. *ISIJ Int.* **2000**, *40*, 121–128. [CrossRef]
17. Zhang, L. Several important scientific research points of non-metallic inclusions in steel. *Steelmaking* **2016**, *32*, 1–16.
18. Alhussein, A.; Yang, W.; Zhang, L. Effect of interactions between Fe–Al alloy and MgO-based refractory on the generation of $MgO{\cdot}Al_2O_3$ spinel. *Ironmak. Steelmak.* **2020**, *47*, 424–431. [CrossRef]
19. Huang, F.; Zhang, L.; Zhang, Y.; Ren, Y. Kinetic modeling for the dissolution of MgO lining refractory in Al-killed steels. *Metall. Mater. Trans. B* **2017**, *48*, 2195–2206. [CrossRef]
20. Wang, X.; Li, X.; Li, Q.; Huang, F.; Li, H.; Yang, J. Control of stringer shaped non-metallic inclusions of CaO-Al_2O_3 System in API X80 linepipe steel plates. *Steel Res. Int.* **2014**, *85*, 155–163. [CrossRef]
21. Deng, Z.; Zhu, M. Evolution mechanism of non-metallic inclusions in Al-killed alloyed steel during secondary refining process. *ISIJ Int.* **2013**, *53*, 450–458. [CrossRef]
22. Jing, G.; Shu-Sen, C.; Zi-Jian, C. Mechanism of non-metallic inclusion formation and modification and their deformation during compact strip production (CSP) process for aluminum-killed steel. *ISIJ Int.* **2013**, *53*, 2142–2151. [CrossRef]
23. Jiang, M.; Wang, X.; Chen, B.; Wang, W. Laboratory study on evolution mechanisms of non-metallic inclusions in high strength alloyed steel refined by high basicity slag. *ISIJ Int.* **2010**, *50*, 95–104. [CrossRef]
24. Liu, C.; Gao, X.; Ueda, S.; Kitamura, S. Change in composition of inclusions through the reaction between Al-killed steel and the slag of CaO and MgO saturation. *ISIJ Int.* **2019**, *59*, 268–276. [CrossRef]
25. Strandh, J.; Nakajima, K.; Eriksson, R.; Jönsson, P. Solid inclusion transfer at a steel-slag interface with focus on tundish conditions. *ISIJ Int.* **2005**, *45*, 1597–1606. [CrossRef]
26. Strandh, J.; Nakajima, K.; Eriksson, R.; Jönsson, P. A mathematical model to study liquid inclusion behavior at the steel-slag interface. *ISIJ Int.* **2005**, *45*, 1838–1847. [CrossRef]
27. Shannon, G.N.; Sridhar, S. Modeling Al_2O_3 inclusion separation across steel-slag interfaces. *Scand. J. Metall.* **2005**, *34*, 353–362. [CrossRef]
28. Xuan, C.; Persson, E.S.; Sevastopolev, R.; Nzotta, M. Motion and detachment behaviors of liquid inclusion at molten steel–slag interfaces. *Metall. Mater. Trans. B* **2019**, *50*, 1957–1973. [CrossRef]
29. Liu, C.; Yang, S.; Li, J.; Zhu, L.; Li, X. Motion behavior of nonmetallic inclusions at the interface of steel and slag. Part I: Model development, validation, and preliminary analysis. *Metall. Mater. Trans. B* **2016**, *47*, 1882–1892. [CrossRef]
30. Liu, W.; Yang, S.; Li, J.; Wang, F.; Yang, H. Numerical model of inclusion separation from liquid metal with consideration of dissolution in slag. *J. Iron Steel Res. Int.* **2019**, *26*, 1147–1153. [CrossRef]
31. Lee, S.H.; Tse, C.; Yi, K.W.; Misra, P.; Chevrier, V.; Orrling, C.; Sridhar, S.; Cramb, A.W. Separation and dissolution of Al_2O_3 inclusions at slag/metal interfaces. *J. Non-Cryst. Solids* **2001**, *282*, 41–48. [CrossRef]
32. Liu, C.; Jia, Y.; Hao, L.; Han, S.; Huang, F.; Yu, H.; Gao, X.; Ueda, S.; Kitamura, S. Effects of slag composition and impurities of alloys on the inclusion transformation during industrial ladle furnace refining. *Metals* **2021**, *11*, 763. [CrossRef]
33. Liu, L.; Fan, J.; Wang, P.; Wang, L. Generation mechanism of large inclusions during bearing steels refining process by tracer method. *Iron Steel* **2017**, *52*, 34–41.
34. Xu, J.; Wang, K.; Wang, Y.; Qu, Z.; Tu, X.; Meng, X. Influence mechanism of silicon content in al-killed steel on compositions of inclusions during LF refining. *Ironmak. Steelmak.* **2021**, *48*, 127–132. [CrossRef]
35. Kang, D.; Yu, Y. Determination of Content of Trace Magnesium and Calcium in Iron and Steel by Inductively Coupled Plasma Mass Spectrometry. *Angang Technol.* **2013**, *5*, 37–40.
36. Howie, R.A.; Zussman, J.; Deer, W. *An Introduction to the Rock-Forming Minerals*; Longman: London, UK, 1992.

metals

Article

Study of Process Parameters on Solidification Structure and Centre Grain Size of 2311 in 420 mm Extra-Thick Continuously Cast Slabs

Lijun Xu [1,2,*], Pan Zhang [1,*], Yong Shuai [3], Pengzhao Shi [1], Zhonghua Zhan [4] and Minglin Wang [1]

1 National Engineering Research Center of Continuous Casting Technology, Central Iron and Steel Research Institute, Beijing 100081, China
2 Material Digital R&D Center, China Iron & Steel Research Institute Group, Beijing 100081, China
3 Technology Center, Xinyu Iron & Steel Group Co., Ltd., Xinyu 338001, China
4 State Key Laboratory of Advanced Metallurgy, University of Science and Technology Beijing, 30 Xueyuan Road, Beijing 100083, China
* Correspondence: ljxuah@sina.com (L.X.); zhangpan19980203@163.com (P.Z.)

Abstract: Based on the solidification heat transfer model and the CAFE model, the solidification behavior and structure of 2311 die steel, with a cross-section dimension of 415 × 2270 mm at different casting speeds, specific water flow and superheat, is numerically simulated. Nail-shooting and acid-etching experiments are carried out on the slab to verify the model's macroscopic size. With the increase in casting speed, the slab's central equiaxed grain ratio (ECR) decreases and the average grain size increases. The increase in superheat promotes the growth of columnar grains and inhibits the growth of central equiaxed grains. When the superheat increases from 23 to 38 K, the ECR decreases from 43.2 to 29.64%, and the average radius of grains increases from 0.89 to 1.01 mm. With the increase in specific water flow, the ECR decreases, and the average grain radius is the smallest when the specific water content is 0.32 L kg^{-1}. Finally, the slab quality is improved by process optimization, and the central segregation index of carbon decreases from mean value of 1.15 to 1.05.

Keywords: CAFE; equiaxed crystal ratio; macrosegregation; numerical simulation; continuous casting; process optimization

Citation: Xu, L.; Zhang, P.; Shuai, Y.; Shi, P.; Zhan, Z.; Wang, M. Study of Process Parameters on Solidification Structure and Centre Grain Size of 2311 in 420 mm Extra-Thick Continuously Cast Slabs. *Metals* 2023, 13, 47. https://doi.org/10.3390/met13010047

Academic Editor: Noé Cheung

Received: 5 November 2022
Revised: 19 December 2022
Accepted: 21 December 2022
Published: 24 December 2022

1. Introduction

As a widely used plastic die steel in the world, 2311 die steel has good anti-rust ability. With improvement of the internal quality control ability of the continuous casting slab, the continuous casting process has become the main method to prepare 2311 die steel, but the center segregation of 2311 in 420 mm extra-thick is very serious [1]. This is mainly because the extra-thick slab has large thickness, low casting speed and long solidification time, which makes it easy to produce central segregation in the production process [2,3]. Soft reduction and electromagnetic stirring are main means to control the macrosegregation of the casting slab [4,5], which can compensate for shrinkage at the end of solidification and increase the ECR [6]. Soft reduction technology can obviously improve the center quality of the slab and effectively improve the quality of the subsequent rolled products [7–9].

The soft reduction technique was developed in the 1990s and is considered the best way to minimize centerline segregation by reducing the shrinkage cavity to compensate for negative pressure [10]. Rogberg et al. studied the effect of reduction amount on centerline segregation in high carbon and stainless steel [11]. Zhu et al. studied the influence of reduction position and reduction on center segregation by numerical simulation [10]. VAI believes that when the f_s < 30 %, the central porosity and segregation almost does not occur. At this time, the shrinkage roll gap makes it easy to increase the incidence of internal cracks, so it is recommended that the best solid phase rate (f_s) is 0.30–0.95 [12]. Chen et al. believed that the central segregation was significantly improved by the soft reduction when the f_s is between 0.55 and 0.75 [13]. Moreover, with the increase in slab thickness, it is difficult to find the best optimization parameters. Slabs with a thickness of 210,240 and 300 mm are the most studied. However, the influence of reduction parameters on macrosegregation were rarely studied for 420 mm thickness. Only a few people have studied the optimum parameters for 400 mm thickness [14]. In actual production, the continuous casting process parameters will affect the solidification end position, thus affecting the reduction position. Therefore, it is particularly important to determine the soft reduction parameters by studying the influence of the process parameters on the solidification structure and the position of the solidification end point of the 420 mm extra-thick slab. To determine the solidification end point, it is necessary to measure the solidified shell thickness of the continuous casting slab. Because a nail-shooting test is convenient and accurate [15], the nail-shooting method is commonly used to determine the solidified shell thickness of continuous casting slabs.

In addition, macrosegregation is also closely related to the solidification characteristic structure [16,17]. This is mainly because macrosegregation is closely related to the liquid phase flow between dendrites [18–20], and grain size and the ECR will affect the liquid phase flow between dendrites [21–23]. Therefore, it is valuable to study the grain size and ECR of the continuous casting slab. Although research has provided some useful information for practical production, there are few studies on the simulation of the solidification structure of ultra-thick slabs.

In this paper, the solidification heat transfer model of a 2311 die steel slab with an area of 2270 × 415 mm^2 is established by ProCAST software, and the accuracy of the model is verified by a nail-shooting experiment and surface temperature measurement. Then, the solidification structure of the slab is simulated and verified by an acid etching test. The ECR and the average grain size under the different casting speeds, superheats and specific water are calculated and analyzed. On this basis, the soft reduction position and reduction amount are optimized. Finally, an industrial test is carried out to verify the optimized process parameters, which effectively improves the central segregation of the slab.

2. Model Descriptions

2.1. Solidification Heat Transfer Model

To simplify the calculation process of the numerical simulation, the following assumptions are made [24]:

1. Heat transfer along the casting direction is ignored.
2. The heat exchange of the slab in the secondary cooling zone is described by the heat transfer coefficient.
3. The influence of mold vibration and protective slag film on heat transfer is ignored.
4. The effect of fluid flow on heat transfer is described by adjusting the thermal conductivity.

Equations (1) and (2) are the two-dimensional heat transfer-governing equations for slab continuous casting:

$$\frac{\partial}{\partial x}\left(\lambda\frac{\partial T}{\partial x}\right)+\frac{\partial}{\partial y}\left(\lambda\frac{\partial T}{\partial y}\right)=\rho\frac{\partial H}{\partial T} \quad (1)$$

$$H=\int_0^T C_P dT+L\,(1-f_s) \quad (2)$$

where T is the temperature, t is the time, ρ is the liquid density, C_P is the heat capacity, λ is the effective thermal conductivity, L is the latent heat of fusion and f_s is the solid fraction. The main chemical compositions of 2311 die steel are listed in Table 1. The composition of steel was inputted into the material database to calculate the density, enthalpy, solid fraction and other thermal physical parameters.

Table 1. Chemical composition (mass%) of 2311 die steel.

Fe	C	Si	Mn	P	S	Cr	Al	Ni
95.4	0.398	0.2881	1.43	0.0178	0.0018	1.89	0.02	0.0155

2.2. Nucleation Model

In 1989, Rappaz [25] proposed a continuous nucleation model based on Gaussian distribution, as shown in Equations (3) and (4):

$$n(\Delta T) = \int_0^{\Delta T} \frac{dn}{d(\Delta T)} d(\Delta T) \tag{3}$$

$$\frac{dn}{d(\Delta T)} = \frac{n_{max}}{\sqrt{2\pi}\Delta T_\sigma} exp[-\frac{(\Delta T - \Delta T_n)^2}{\Delta T_\sigma{}^2}] \tag{4}$$

where ΔT is the undercooling, K; n_{max} is the maximum nucleation density, m^{-3} for volume and m^{-2} for surface; ΔT_σ is the standard deviation undercooling, K; and ΔT_n is the mean undercooling, K. The effect of melt movement on the grain morphology cannot be considered in ProCAST software. The effect of melt movement on the grain morphology can only be expressed by adjusting nucleation parameters. The effect of fluid flow on the equiaxed grain rate is described by increasing the number of nuclei, but the effect of melt movement on the grain morphology cannot be described directly. The nucleation parameters and growth parameters used in this paper are shown in Table 2.

Table 2. Nucleation parameters of CAFE model.

Parameters	$\Delta T_{S,max}$, K	$\Delta T_{S,\sigma}$, K	n_S	$\Delta T_{V,max}$, K	$\Delta T_{V,\sigma}$, K	n_V
Values	1	0.1	2.5×10^6	2.5	1	3.0×10^9

Note: $n_V = 0.8\, n_S{}^{3/2}$.

2.3. Dendrite Tip Growth Kinetics Model

In casting, the growth kinetics of equiaxed crystals and columnar crystals can be calculated by the KGT model [26,27]. Based on the marginal stability criterion, Equation (5) is obtained:

$$V^2 \frac{\pi^2 \Gamma}{P^2 D^2} + V \frac{mC_0(1-k_0)}{D\,[1-(1-k_0)Iv(P)]} + G = 0 \tag{5}$$

where V is the dendritic tip growth rate, $m \cdot s^{-1}$; Γ is the Gibbs-Thomson coefficient, P is the solutal Peclet number, D is the solute diffusion coefficient, m is the liquidus slope, C_0 is the nominal concentration, k_0 is the partition coefficient, G is the temperature gradient, and set as 0, without considering the thermal effect, $Iv(P)$ is the Ivantsov function $Iv(P) - Pexp(P)E_1(P)$ and E_1 is the exponential integral.

In the actual calculation process, to accelerate the calculation process, the KGT model is fitted to obtain the following equation (Equations (6)–(8)) [28]:

$$V(\Delta T) = \alpha_2 \Delta T^2 + \alpha_3 \Delta T^2 \tag{6}$$

$$\alpha_2 = [\frac{-\rho}{2mC_0(1-k)^2 \Gamma k} + \frac{1}{mC_0(1-k)D}] \frac{D^2}{\pi^2 \Gamma} \tag{7}$$

$$\alpha_3 = \frac{D}{\pi \Gamma} \cdot \frac{1}{(mC_0)^2(1-k)} \tag{8}$$

where α_2 and α_3 are the fitting polynomial coefficients of dendrite tip growth kinetics parameters, the units are $m \cdot s^{-1} \cdot K^{-2}$ and $m \cdot s^{-1} \cdot K^{-3}$ and ρ is the density of steel.

In this paper, the 2311 die steel is divided into seven binary systems of Fe-C, Fe-Si, Fe-Mn, Fe-P, Fe-S, Fe-Al and Fe-Cr. The composition c, distribution coefficient k, liquidus slope m, solute diffusion coefficient D [29] and Gibbs-Thompson coefficient Γ [30] of Fe-X alloys are shown in Table 3.

Table 3. Liquidus slope, solute partition coefficient, diffusion coefficient and Gibbs-Thompson coefficient of binary Fe-X steels.

Element	K	m/K·(wt.%)$^{-1}$	D/m^2·s^{-1}	Γ/m·K
C	0.16	−87.14	1.1×10^{-8}	
Si	0.61	−44.93	3.5×10^{-9}	
Mn	0.71	−5.32	2.4×10^{-9}	
P	0.25	−32.39	1.9×10^{-9}	3×10^{-7}
S	0.06	−44.93	3.9×10^{-9}	
Cr	0.9	−1.73	3.5×10^{-9}	
Al	1.22	−4.9	3.0×10^{-9}	

3. Solution Conditions and Model Validation

3.1. Geometric Model

The solidification process of the slab is simulated by the thin−slicing method. The ProCAST software is used for grid division. The geometric model and mesh are shown in Figure 1.

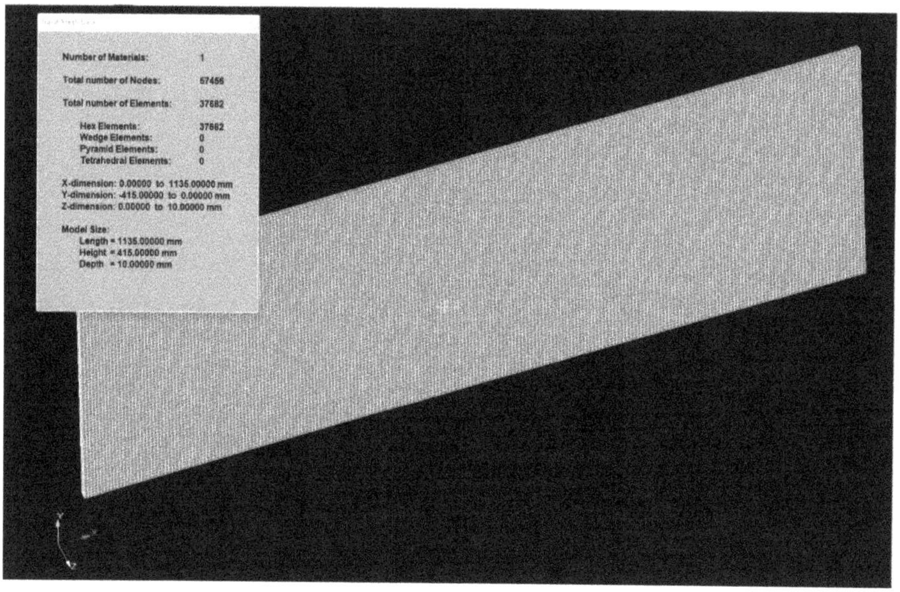

Figure 1. Information about geometric model.

3.2. Initial and Boundary Conditions

3.2.1. Initial Condition

The initial condition is given by Equation (9):

$$T_0 = T_C \tag{9}$$

where T_0 is the initial temperature, K; T_C is the casting temperature, K.

3.2.2. Boundary Condition

In the mold, the heat flux is calculated by Equations (10) and (11) [31]:

$$q_m = \left(2.688 - \beta\sqrt{t}\right) \times 10^6 \tag{10}$$

$$\beta = \frac{1.5\times(2.688\times10^6 - q_c)}{\sqrt{l_m/v}} \tag{11}$$

where q_m is the heat flux of the mold, $W\cdot m^{-2}$; β is a coefficient depending on the mold cooling condition, $W\cdot m^{-2}\cdot s^{-1/2}$; q_c is the average heat flux in the mold, $W\cdot m^{-2}$; l_m is the effective length of the mold, m; t is the time in the mold, s and v is the casting speed, $m\cdot s^{-1}$.

In the foot roller section and the secondary cooling zone, the heat transfer coefficient is calculated by Equations (12) and (13):

$$h_1 = \alpha\cdot[581W^{0.541}(1 - 0.0075T_W) \tag{12}$$

$$h_2 = \gamma\cdot(130 + 350W) \tag{13}$$

where h_1 and h_2 are the heat transfer coefficient of the foot roller section and the secondary cooling zone, $W/(m^2\cdot K)$; W is the water flow rate, $L/(m^2\cdot s)$; T_w is the temperature of the environment, K and α and γ are the correction factors.

In the air-cooling zone, the heat flux is calculated by Equations (14) and (15) [32]:

$$q_a = \varepsilon\sigma(T^4 - T_a{}^4) \tag{14}$$

$$\varepsilon = 0.85/(1 + \exp(42.68 - 0.02682T_s)^{0.0115} \tag{15}$$

where ε is the radiation coefficient and σ is Stefan-Boltzmann constant, 5.67×10^{-8} $W/(m^2\cdot K^4)$.

The cooling parameters are shown in Table 4.

Table 4. Parameters at different sections of secondary cooling zone.

Secondary Cooling Zone	Length, m	Distance from the Meniscus, m	Water Flow Rate, $L\cdot min^{-1}$
Foot roller section (W)	0.29	0.8–1.09	93.4
Foot roller section (N)	0.89	/	161.9
L2	1.68	1.09–2.77	188.2
L3	1.77	2.77–4.54	84.1
L4	1.87	4.54–6.41	75.0
L5	2.13	6.41–8.54	73.8
L6	2.13	8.54–10.64	35.6
L7	4.49	10.67–15.16	71.4
L8	4.71	15.16–19.87	68.9
L9	2.36	19.87–22.23	37.6
L10	5.16	22.23–27.39	63.6
L11	4.75	27.39–32.14	60.2
L12	5.00	32.14–37.61	53.2
L13	7.00	37.61–44.91	64.2

3.3. Model Validation

3.3.1. Validation of Heat Transfer

The surface temperature measured at half of the narrow surface of the slab is compared with the simulation results, as shown in Figure 2. The error between the simulated temperature and the measured temperature is within 5%. The absolute error is not more than 40 °C, which indicates that the processing of model boundary conditions is close to the actual production process. In this paper, the accuracy of the solidification heat transfer model is further verified by the nail-shooting experiment. The liquid zone thickness at the end of segment 7 is 84 mm and the actual shell thickness is 331 mm, as shown in Figure 3a. The liquid zone thickness at the end of segment 8 is 56 mm and the actual shell thickness is 359 mm, as shown in Figure 3b, and the simulated shell thickness is 351.2 mm, with an error of 2.2%. The liquid core thickness at the end of segment 9 is 17 mm and the actual shell thickness is 394 mm, as shown in Figure 3c. The solidification end position is 27.54 m from the meniscus. The error does not exceed 2.2%, which accurately reflects the solidification characteristics of the slab continuous casting process.

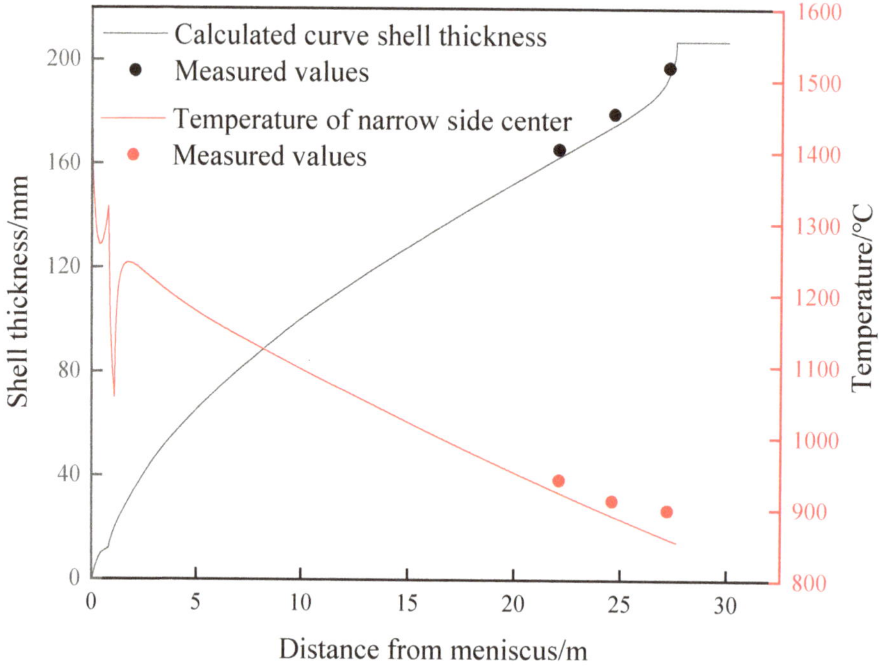

Figure 2. Comparison of the calculated temperature of the slab narrow side center.

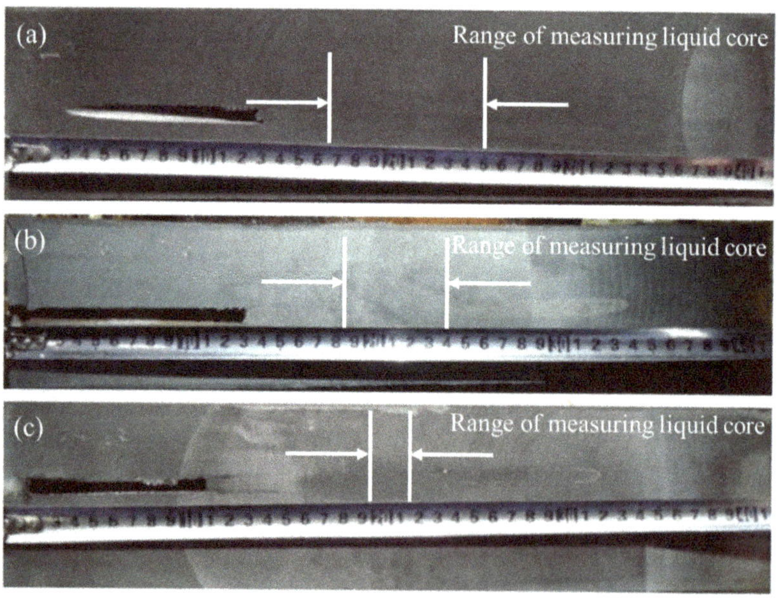

Figure 3. Photographs of sulfur printing samples for nail-shooting experiments: (**a**) the end of segment 7; (**b**) the end of segment 8; (**c**) the end of segment 9. The white arrows represent the range of liquid core.

3.3.2. Validation of Solidification Structure

The calculated ECR is 29.64%, as shown in Figure 4a. The macrograph of the solidification structure after pickling is shown in Figure 4b, and the ECR in the inner arc is about 30.36%. Different colors in Figure 4 represent grain orientation. The experimental and simulated ECR are almost identical, which indicates the model accuracy is high enough to predicate the solidification structure of the slab.

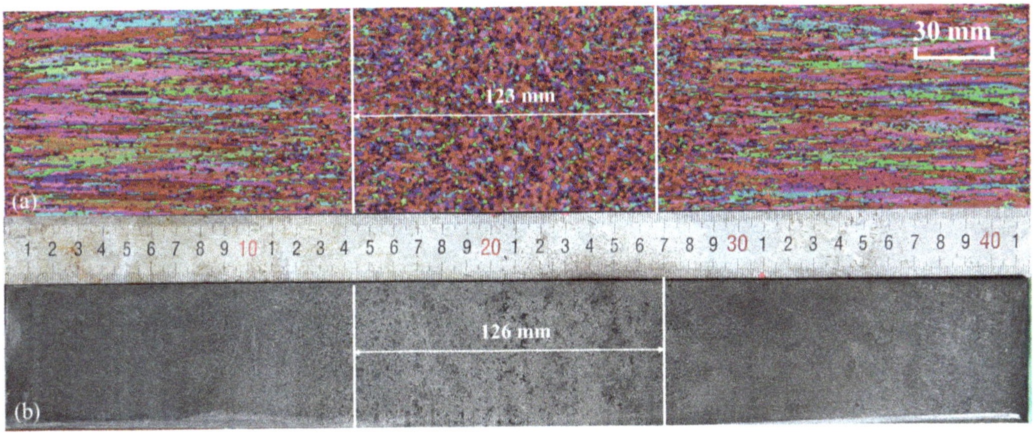

Figure 4. Comparison of numerical and experimental structure of 2311 die steel: (**a**) simulated solidification structure (**b**) experimental solidification structure.

4. Results and Discussion

4.1. Effect of Superheat on Solidification Structure

Under the casting speed of 0.5 m·min^{-1} and the specific water flow of 0.32 L·kg^{-1}, the solidification structures with various superheats of 23, 28, 33 and 38 K are simulated.

When the superheat is 23 K, the ECR is 43.2%, as shown in Figure 5a. The ECR is 37.84% when the superheat is 28 K, as shown in Figure 5b. As can be seen from Figure 5c, the ECR is 32.43%. Figure 5d shows the solidification structure when the superheat is 38 K and the ECR is 29.64%. With the increase in superheat, the ECR decreases and the columnar crystals zone expands. At the same time, the width of the columnar crystals increases significantly. This is because with the increase in superheat, the number of free grains decreases, which leads to the decrease in the number of nucleation sites at the solid–liquid interface and the decrease in undercooling degree. This promotes the growth of dendrites and inhibits the development of equiaxed crystals. In addition, the excessive growth of dendrites leads to the widening of the columnar crystal zone. This may lead to the formation of a bridge, making shrinkage more serious. Too low superheat can cause nozzle clogging. Therefore, it is suggested that the appropriate superheat should be adopted to obtain fine grains and high ECR.

Table 5 lists the statistical results of grain parameters and distribution under different superheats. When the superheat is 23 K, the total number of grains is 20,509, and when the superheat rises to 38 K, the total number of grains is 17,025. When the superheat increases from 23 to 38 K, the average surface area of the grains increases from 0.8260 × 10^{-6} to 0.9950 × 10^{-6} m^2, and the average radius of the grains increases from 0.89051 to 1.01031 mm. With the increase in superheat, the number of grains per meter and per square meter decreases.

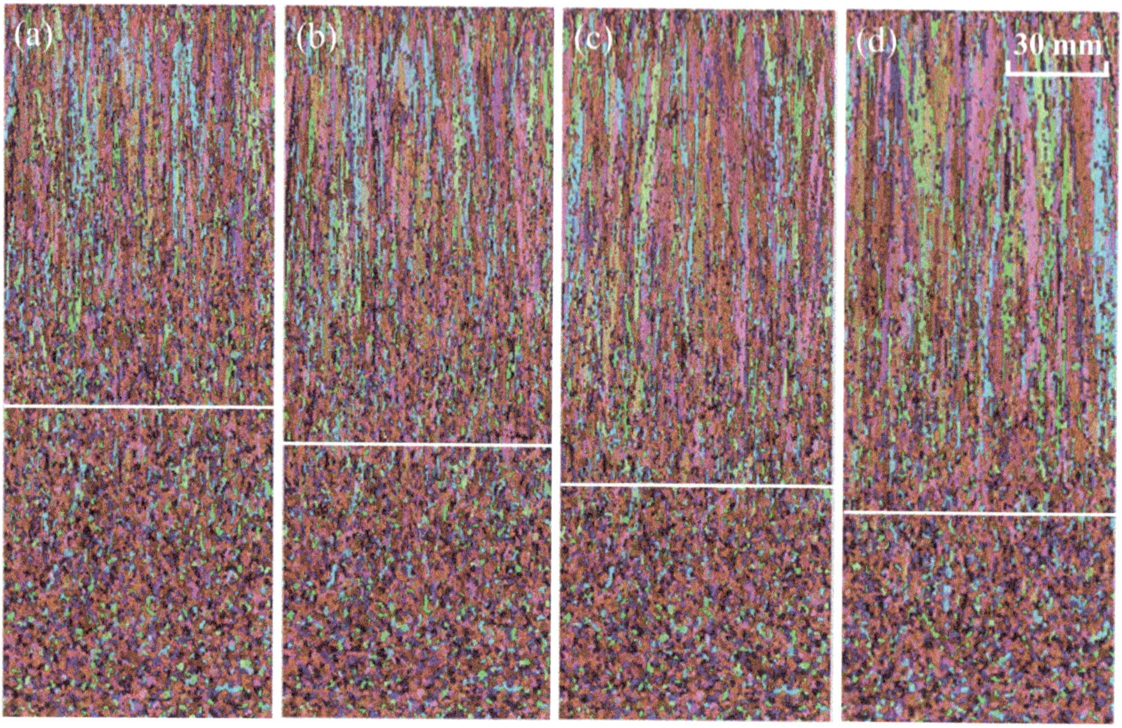

Figure 5. Simulated solidification structures under different superheats: (**a**) 23 K; (**b**) 28 K; (**c**) 33 K; (**d**) 38 K. (Different colors in figure represent grain orientation).

Table 5. Statistical results of grain parameters and distribution in simulated area under different superheats.

ΔT	No. Grains	Mean Surf/m^2	Nl/m	Na/m^2	Mean Radius/m
23	20,509	0.8260×10^{-6}	1.6934921×10^{3}	1.2106648×10^{6}	0.89051×10^{-3}
28	19,239	0.8802×10^{-6}	1.6720469×10^{3}	1.1356956×10^{6}	0.93727×10^{-3}
33	18,061	0.9379×10^{-6}	1.6245649×10^{3}	1.0664572×10^{6}	0.97005×10^{-3}
38	17,025	0.9950×10^{-6}	1.5949357×10^{3}	1.0050000×10^{6}	1.01031×10^{-3}

4.2. Effect of Casting Speed on Solidification Structure

With a superheat of 38 K and a specific water flow of 0.32 L·kg^{-1}, the solidification structures with casting speeds of 0.45, 0.50, 0.55 and 0.60 m·min^{-1} are simulated. When the casting speed is 0.45, 0.50, 0.55 and 0.60 m min^{-1}, the ECR is 31.08 (Figure 6a), 29.64 (Figure 6b), 28.67 (Figure 6c) and 27.71% (Figure 6d), respectively.

Under the same specific water flow, the ECR decreases with the increase in casting speed. In the range of 0−15 mm from the slab surface, the cooling rate decreases with the increase in casting speed, as shown in Figure 7a. The main reason is that with the increase in casting speed, the residence time of molten steel in the mold decreases, which makes the export of superheat delayed. This promotes the growth of columnar crystals and reduces the ECR. In the range of 15−50 mm from the slab surface, the cooling rate increases with the increase in casting speed, as shown in Figure 7b. This is mainly due to being under the same specific water; the greater the total water, the higher the cooling intensity, resulting in a bigger temperature gradient at the solid–liquid interface, thereby promoting the growth of columnar crystals. The cooling rate is basically the same in the range of 50−220 mm from the slab surface, as shown in Figure 7c,d. The cooling water volume is small, so that there is no significant difference in cooling intensity. As the casting speed increases, the average grain radius increases, as shown in Figure 8. The reason for this is the short residence time of the slab in the mold and the secondary cooling zone, which is unfavorable for eliminating the superheat of molten steel.

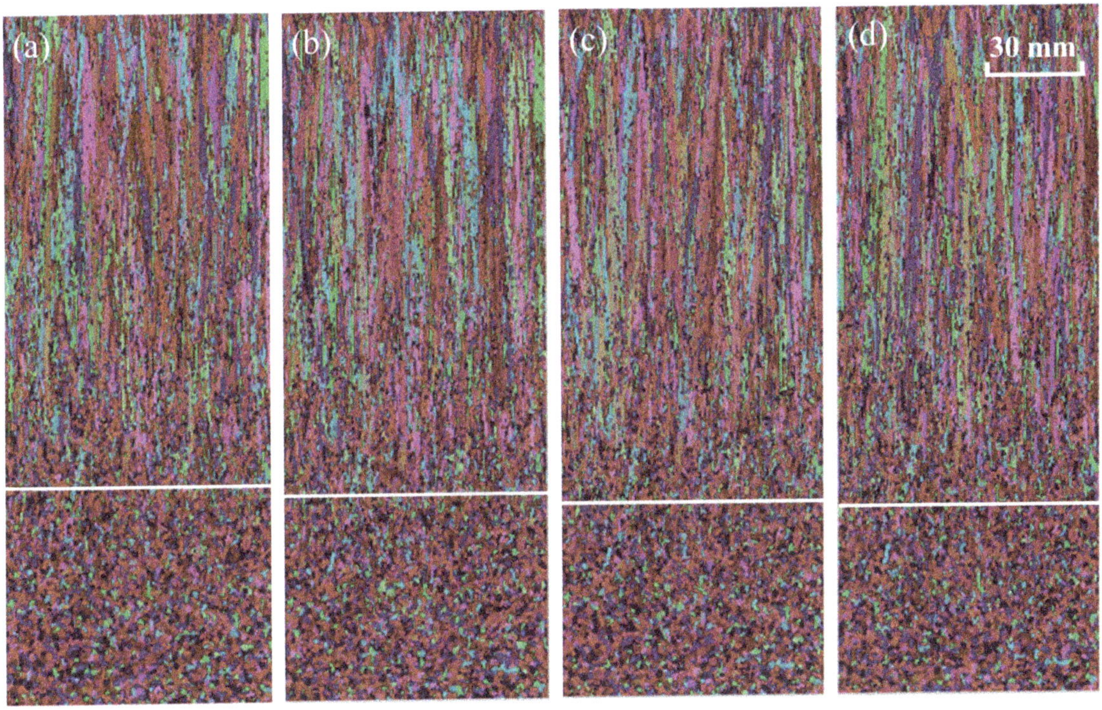

Figure 6. Simulated solidification structures under different casting speeds: (**a**) 0.45 m·min^{-1}; (**b**) 0.45 m·min^{-1}; (**c**) 0.50 m·min^{-1}; (**d**) 0.50 m·min^{-1}. (Different colors in figure represent grain orientation).

4.3. Effect of Specific Water Flow on Solidification Structure

Four kinds of specific water flow of 0.28, 0.32, 0.36 and 0.40 L·kg^{-1} were selected to study the effect of specific water flow on the surface temperature of slab. With the increase in specific water flow, the liquid core length of the slab decreases. For every 0.04 L·kg^{-1} increase in specific water flow, the position which superheat is eliminated is decreased by about 0.25 m, as shown in Figure 9a. Before the slab completely solidified, the specific water flow has a great influence on the slab surface temperature, especially in the secondary cooling zone, as shown in Figure 9b. For every 0.04 L·kg^{-1} increase in specific water flow, the surface temperature of the slab decreases by about 20 °C. As the specific water flow increases from 0.28 to 0.40 L·kg^{-1}, the superheat−eliminating position and the solidification position slightly reduces.

With a superheat of 38 K and the casting speed of 0.5 m·min^{-1}, the solidification structures with the specific water flow of 0.28, 0.32, 0.36, and 0.40 L·kg^{-1} are simulated. When the specific water flow is 0.28 L·kg^{-1}, the ECR is 31.08%, as shown in Figure 10a. The ECR is 29.64% when the specific water flow is 0.32 L·kg^{-1}, as shown in Figure 10b. It can be seen from the Figure 10c, the ECR is 28.67%. Figure 10d shows the solidification structure when the specific water flow is 0.40 L·kg^{-1}, the ECR is 27.71%. It indicates that as the specific water flow increases, the ECR decreases. The main reason is that with the increase in specific water, the heat transfer coefficient in the secondary cooling zone also increases, which makes the cooling intensity increase. The increase in temperature gradient promotes the growth of columnar crystals and enlarges the columnar zone.

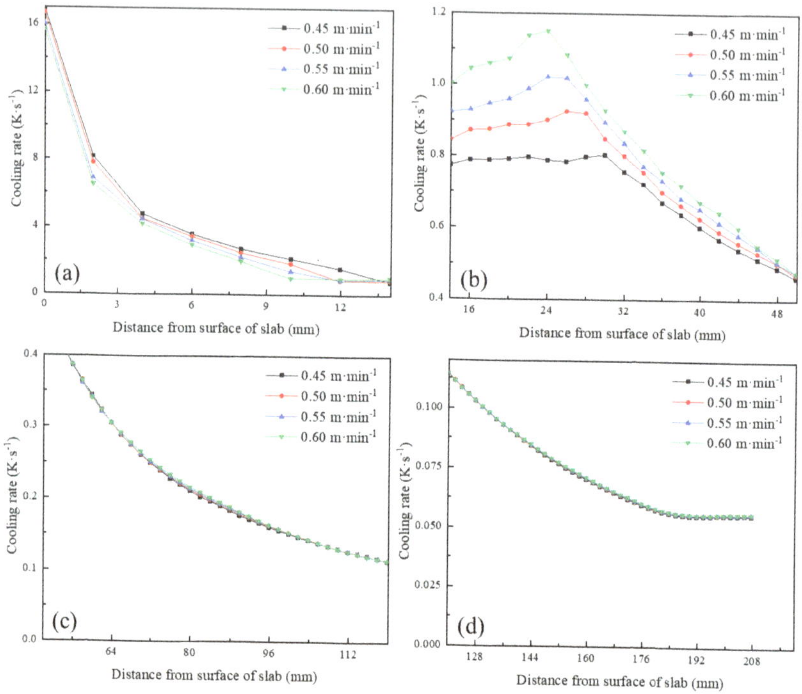

Figure 7. Cooling rate at different positions from surface of slab under different casting speeds: (**a**) 0–15 mm; (**b**) 15–50 mm; (**c**) 50–120 mm; (**d**) 120–220 mm.

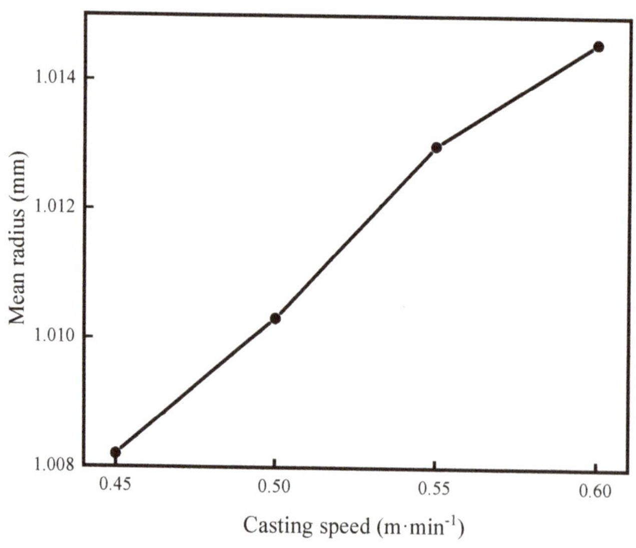

Figure 8. The average grain size of the slab with different casting speeds.

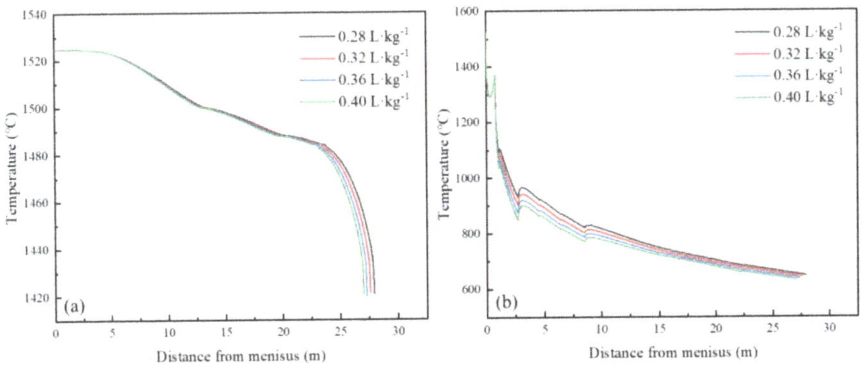

Figure 9. Surface and center temperature curves of casting slab under different specific water flow: (**a**) center temperature; (**b**) surface temperature.

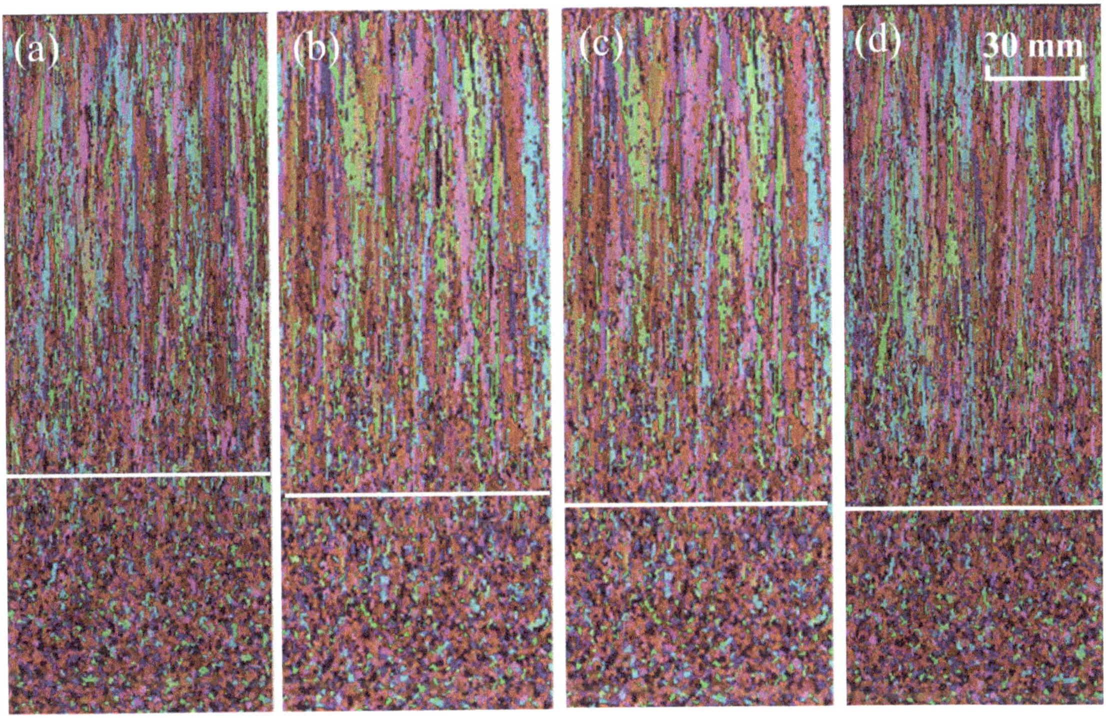

Figure 10. Simulated solidification structures under different specific water flow: (**a**) 0.28 L·kg^{-1}; (**b**) 0.32 L·kg^{-1}; (**c**) 0.36 L·kg^{-1}; (**d**) 0.40 L·kg^{-1}. (Different colors in figure represent grain orientation).

As the specific water flow increases from 0.28 to 0.40 L·kg^{-1}, the average grain radius decreases first and then increases, as shown in Figure 11. When the specific water flow is 0.32 L·kg^{-1}, the average grain radius reaches the minimum. With the increase in specific water flow, the cooling intensity increases, which makes the grain refined. In addition, the slab surface temperature is low (Figure 9b) and the temperature gradient in the thickness direction is large, which promotes the growth of columnar crystals. When the specific water is greater than 0.32 L·kg^{-1}, the cooling intensity is further enhanced. This promotes the growth of columnar crystals; the columnar crystals become wider, and the refined equiaxed crystals are not enough to offset the increase in grain size caused by the expansion of the columnar crystal region. Therefore, the grain size increases.

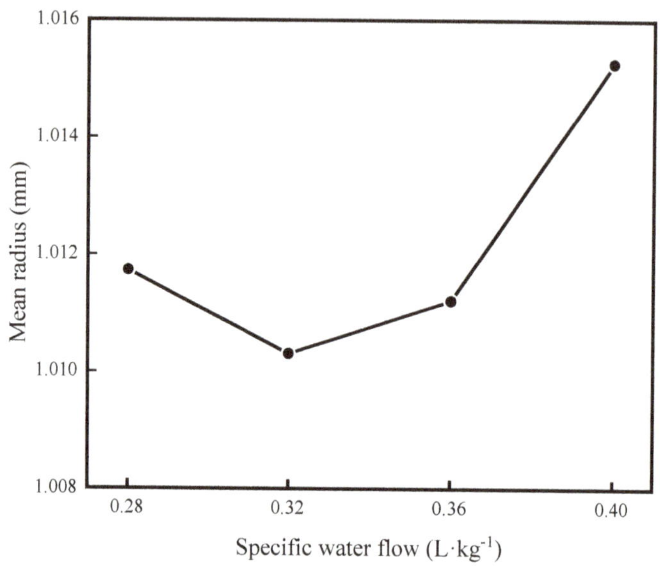

Figure 11. The average grain size of the slab with different specific water flow.

4.4. Optimization of CC Process Parameter

When the specific water flow is 0.32 L/kg, the average grain size reaches the minimum value. Therefore, the specific water flow does not need to be adjusted. According to the simulation results, the increase in casting speed will reduce the ECR and increase the average grain size. Therefore, the adjustment of the process parameters has little effect on the improvement of central segregation.

The center porosity and centerline segregation of the 2311 die slabs are difficult to control. The current control methods mainly include secondary cooling system control, soft/heavy reduction, electromagnetic stirring, etc. The soft reduction is an important method to reduce the center porosity and centerline segregation of the slab. Therefore, the soft reduction process of the slab is optimized. Taking the BT/50 and S355 slab of Xinyu Steel as the research object, the slab center quality was improved by soft reduction when the f_s is between 0.65 and 0.75 [33]. In this paper, when the solid fraction is between 0.7 and 0.8, soft reduction is used to improve the central segregation.

Through the previous calculation and analysis, the slab has completely solidified when it is 27.54 m away from the meniscus. The central solid fraction at the end of segment 7 is less than 0.3. For every 0.04 L·kg^{-1} increase in specific water flow, the position which superheat is eliminated is decreased by about 0.25 m, as shown in Figure 12a. It can be seen from the Figure 12b that the central solid fraction at the end of segment 8 is greater than 0.3. Figure 12c shows the central solid fraction at the end of segment 9. From the position of soft reduction, the solidification end point calculated by the dynamic soft reduction system is relatively forward, and there is no effective soft reduction at the end of solidification, which has little effect on the improvement of central segregation and porosity of the slab. Therefore, it is necessary to move the position of soft reduction backwards (relative to the casting direction) accordingly. The optimized soft reduction parameters are shown in Table 6. The slab before optimization is shown in Figure 13a, showing that center segregation is very serious. Figure 13b shows the slab after optimization.

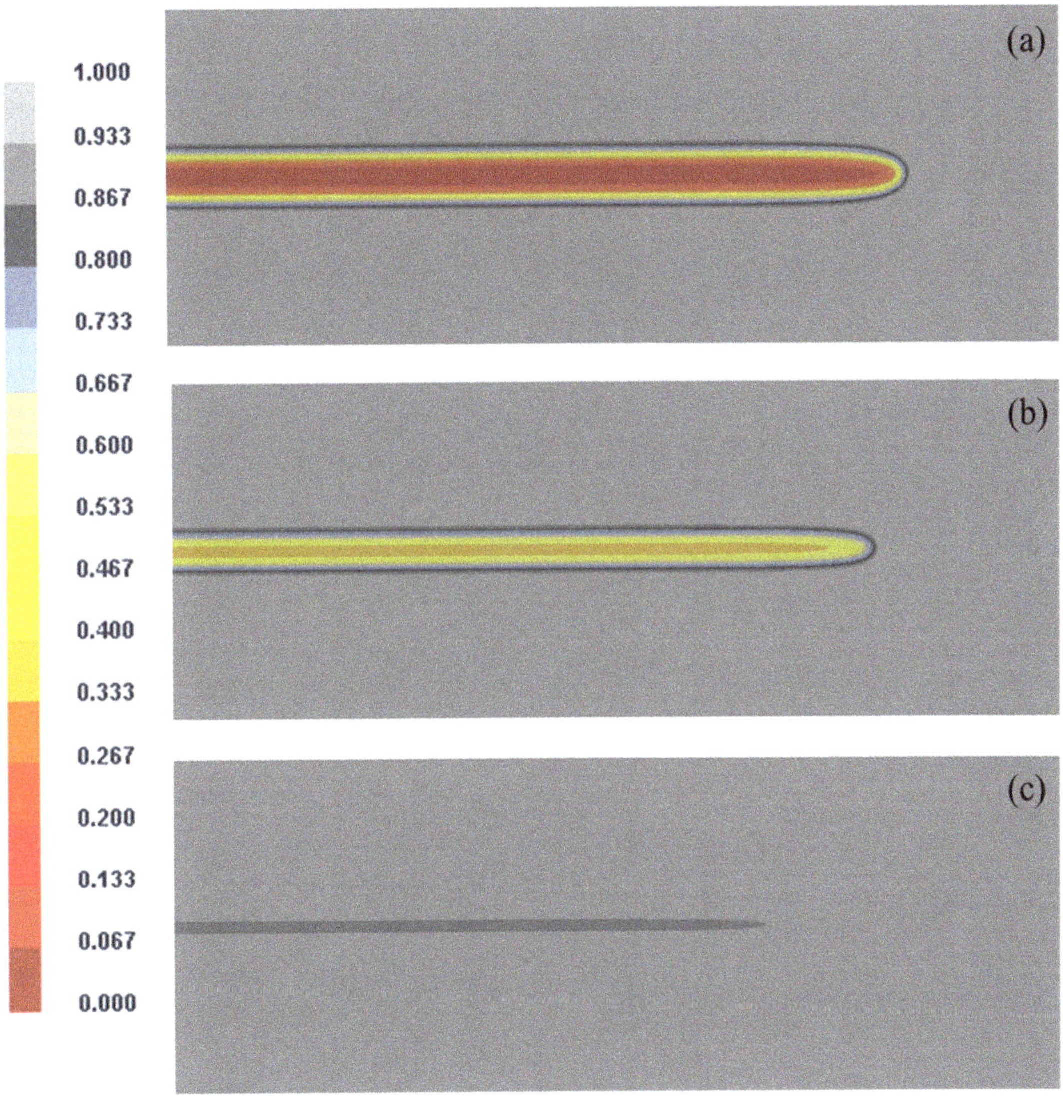

Figure 12. The solid fraction of slab center: (**a**) the end of segment 7; (**b**) the end of segment 8; (**c**) the end of segment 9.

Table 6. The soft reduction parameters after optimization.

Reduction Position	Reduction Amount/mm
Segment 8	4
Segment 9	4
Segment 10	2

To quantitatively describe the macrosegregation, the carbon segregation index of the slabs is quantitatively measured by a chemical analysis of the drillings, as shown in Figure 14. A carbon−sulfur analyzer is used to obtain the carbon content at different locations in the slab. The carbon segregation index before and after optimization is shown in Figure 15. The carbon segregation at the center of the slab is reduced from 1.045 to 0.98. It proves that the centerline segregation is weakened.

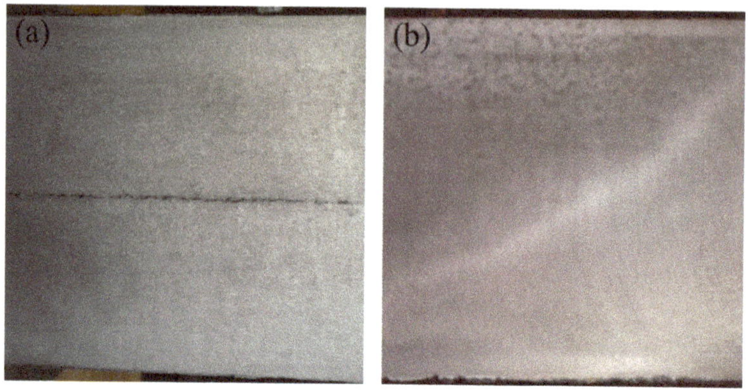

Figure 13. The slab before and after optimization: (**a**) before optimization; (**b**) after optimization.

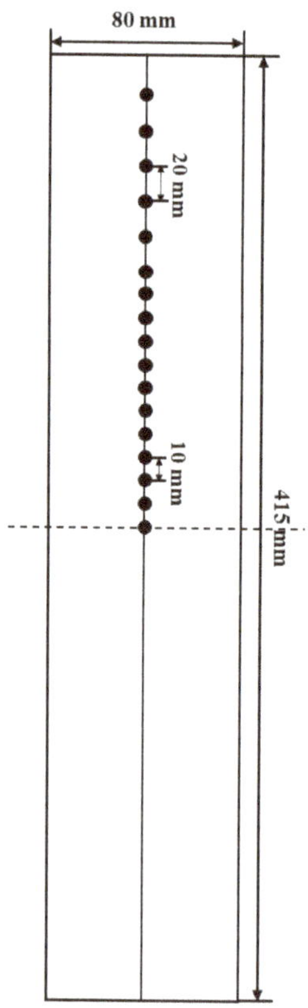

Figure 14. Carbon segregation sampling schematic.

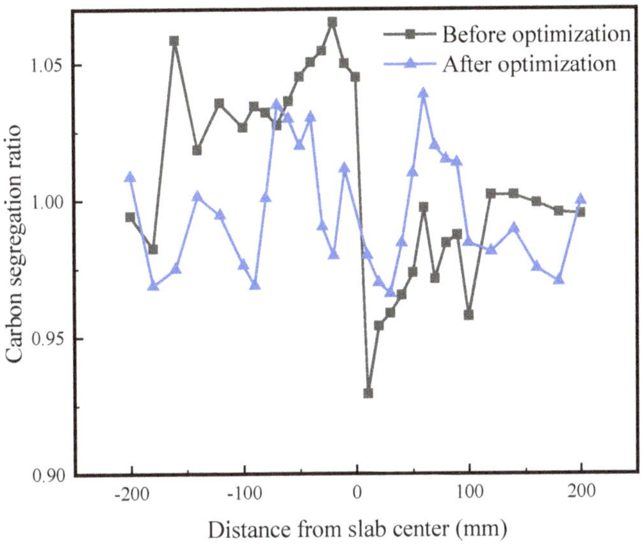

Figure 15. Comparison of carbon segregation index before and after optimization.

5. Conclusions

1. The simulation results are verified by a nail-shooting experiment, acid etching and a surface temperature measurement. The error of the shell thickness and surface temperature is controlled within 5 %.
2. The ECR decreases with the increase in the casting speed and the specific water flow. Superheat has a great influence on the ECR, which is mainly because the nucleation amount in the molten steel decreases with the increase in superheat.
3. The average grain radius increases with the increase in the casting speed and superheat. As the specific water flow increases from 0.28 $L{\cdot}kg^{-1}$ to 0.40 $L{\cdot}kg^{-1}$, the average grain radius decreases first and then increases. When the specific water flow is 0.32 $L{\cdot}kg^{-1}$, the average grain radius reaches the minimum.
4. With the optimized the position of soft reduction, the central carbon segregation is weakened.

Author Contributions: L.X.: original draft preparation, validation; P.Z.: writing—review and editing, Conceptualization; Y.S.: data curation; P.S.: formal analysis; Z.Z.: project administration; M.W.: writing—review and editing, methodology. All authors have read and agreed to the published version of the manuscript.

Funding: This research received no external funding.

Institutional Review Board Statement: Not applicable.

Informed Consent Statement: Not applicable.

Conflicts of Interest: The authors declare no conflict of interest.

References

1. Zhen, X.; Zhu, Z.; Jiang, H.; Li, J.; Wang, Y.; Bai, Z.; Zhao, J.; Sun, B. Effect of the positon on the central segregation of 400 mm ultra-thick slabs. *Steelmaking* **2012**, *28*, 20–23.
2. Miao, J.; Ruan, X.; Han, Z.; Kong, K.; Liu, Q. Control of susceptible quality defects in heavy plate continuous casting process. *Contin. Cast.* **2015**, *40*, 25–28.
3. Ma, R.; Zhen, X. Especially thick slab continuous casting quality defects and the control. *Contin. Cast.* **2011**, *1*, 67–75. [CrossRef]
4. Li, Y.J.; Li, L.; Zhang, J.Q. Study and application of a simplified soft reduction amount model for improved internal quality of continuous casting bloom. *Steel Res. Int.* **2017**, *88*, 170–176. [CrossRef]

5. Jiang, D.B.; Zhu, M.Y. The fluid flow and solidification phenomenon in billet continuous casting process with mold and final electromagnetic stirrings. In *Advances in the Science and Engineering of Casting Solidification*; Nastac, L., Fredriksson, H., Lacaze, J., Hong, C.-P., Catalina, A.V., Buhrig-Polaczek, A., Monroe, C., Sabau, A.S., Ruxanda, R.E.L., Luo, A., et al., Eds.; Springer: Berlin/Heidelberg, Germany, 2015.
6. Li, J.; Wu, H.; Liu, Y.; Sun, Y. Solidification structure simulation and casting process optimization of GCr15 bloom alloy. *China Foundry.* **2022**, *19*, 63–74. [CrossRef]
7. Cenanovic, M.B.; Maureira, H.A.; Ng, M.K.C. Electromagnetic technology for continuous casting in the steel industry. *Direct Roll. Hot Charg. Strand Cast Billets* **1989**, 139–148. [CrossRef]
8. Guo, L.L.; Tian, Y.; Yao, M.; Shen, H.F. Temperature distribution and dynamic control of secondary cooling in slab continuous casting. *Int. J. Miner. Metall. Mater.* **2009**, *16*, 626–631.
9. Yong, C.H.E.N.; Xiao, M.F.; Wu, G.R. Dynamic Soft Reduction Technology for Bloom Casting. *J. Iron Steel Res. Int.* **2010**, *17*, 1–5.
10. Chu, R.-S.; Li, Z.-J.; Liu, J.-G.; Fan, Y.; Liu, Y.; Ma, C.-W. Effect of soft reduction process on segregation of a 400 mm thick high-alloy steel slab. *J. Iron Steel Res. Int.* **2021**, *28*, 272–278. [CrossRef]
11. Rogberg, B.; Ek, L. Influence of Soft Reduction on the Fluid Flow, Porosity and Center Segregation in CC High Carbon- and Stainless Steel Blooms. *ISIJ Int.* **2018**, *58*, 478–487. [CrossRef]
12. Feng, L. *Research on Soft Reduction Theory and Internal Quality Improvement for Continuous Casting Slab*; Northeastern University: Shenyang, China, 2018.
13. Chen, Y.K.; Feng, F. Improvement of center segregation for high carbon steel bloom. In Proceedings of the Seventy Ninth Conference of the Steelmaking Division of the Iron and Steel Society, Pittsburgh, PA, USA, 24–27 March 1996; pp. 505–512.
14. Matsumiya, T. Recent Topics of Research and Development in Continuous Casting. *ISIJ Int.* **2006**, *46*, 1800–1804. [CrossRef]
15. Cherepanov, A.N.; Popov, V.N.; Komshukov, V.P. Flow structure in solidifying continuous-cast steel ingot. *Steel Transl.* **2007**, *37*, 842–846. [CrossRef]
16. Ludlow, V.; Normanton, A.; Anderson, A. Strategy to minimize central segregation in high carbon steel grades during billet casting. *Ironmak. Steelmak.* **2005**, *32*, 68–74. [CrossRef]
17. Ayata, K.; Mori, T.; Fujimoto, T.; Ohnishi, T.; Wakasugi, I. Improvement of macrosegregation in continuously cast bloom and billet by electromagnetic stirring. *Trans. ISIJ* **1984**, *24*, 931–939. [CrossRef]
18. Flemings, M.C.; Nereo, G.E. Macrosegregation: Part I. *Trans. AIME* **1967**, *239*, 1449.
19. Flemings, M.C. Our Understanding of Macrosegregation: Past and Present. *ISIJ Int.* **2000**, *40*, 833–841. [CrossRef]
20. Lesoult, G. Macrosegregation in steel strands and ingots: Characterisation, formation and consequences. *Mater. Sci. Eng. A* **2005**, *19*, 413–414. [CrossRef]
21. Apelian, D.; Flemings, M.C.; Mehrabian, R. Specific permeability of partially solidified dendritic networks of Al-Si alloys. *Metall. Trans.* **1974**, *5*, 2533–2537. [CrossRef]
22. Streat, N.; Weinberg, F. Interdendritic fluid flow in a lead-tin alloy. *Metall. Trans. B* **1976**, *7*, 417. [CrossRef]
23. Poirier, D.R. Permeability for flow of interdendritic liquid in columnar-dendritic alloys. *Metall. Trans. B* **1987**, *18*, 245. [CrossRef]
24. Lally, B.; Biegler, L.; Henein, H. Finite difference heat transfer modeling for continuous casting. *Metall. Mater. Trans. B* **1990**, *21*, 761–770. [CrossRef]
25. Rappaz, M.; Gandin, C.A. Probabilistic modelling of microstructure formation in solidification processes. *Acta Metall. Mater.* **1993**, *41*, 345–360. [CrossRef]
26. Kurz, W.; Giovanola, B.; Trivedi, R. Theory of microstructural development during rapid solidification. *Acta Metallurgica* **1986**, *34*, 823–830. [CrossRef]
27. Kurz, W.; Fisher, D.J. *Fundaments of Solidification*, 4th revised ed.; Trans Tech Publishers: Aedermannsdorf, Switzerland, 1998; p. 77.
28. Bai, L.; Wang, B.; Zhong, H.; Ni, J.; Zhai, Q.; Zhang, J. Experimental and numerical simulations of the solidification process in continuous casting of slab. *Metals* **2016**, *6*, 53. [CrossRef]
29. Li, W.C. *Metallurgy and Physical Chemistry of Materials*; Metallurgical Industry Press: Beijing, China, 2001; pp. 531–533. (In Chinese)
30. Fang, Q.; Ni, H.W.; Zhang, H.; Wang, B.; Lv, Z.A. The effects of a submerged entry nozzle on flow and initial solidification in a continuous casting bloom mold with electromagnetic stirring. *Metals* **2017**, *7*, 146. [CrossRef]
31. Savage, J.; Pritchard, W.H. The problem of rupture of the billet in the continuous casting of steel. *J. Iron Steel Inst.* **1954**, *178*, 269–277.
32. Hardin, R.A.; Liu, K.; Beckermann, C.; Kapoor, A. A transient simulation and dynamic spray cooling control model for continuous steel casting. *Metall. Mater. Trans. B* **2003**, *34*, 297–306. [CrossRef]
33. Shuai, Y. *Solidification Structure Control and Process Optimization of Ultra thick Slab*; Central Iron and Steel Research Institute: Beijing, China, 2020.

metals

MDPI

Article

Modification of Rare Earth Ce on Inclusions in W350 Non-Oriented Silicon Steel

Haijun Wang [1], Yuhao Niu [1], Haitao Ling [1,*], Jialong Qiao [1,2], Yanling Zhang [3], Wei Zhong [4] and Shengtao Qiu [2]

1 Anhui Province Key Laboratory of Metallurgical Engineering & Resources Recycling, Anhui University of Technology, Ma'anshan 243002, China
2 National Engineering Research Center of Continuous Casting Technology, China Iron & Steel Research Institute Group, Beijing 100081, China
3 State Key Laboratory of Advanced Metallurgy, University of Science and Technology Beijing, Beijing 100083, China
4 Xinyu Iron and Steel Co., Ltd., Xinyu 338001, China
* Correspondence: linghaitao@ahut.edu.cn

Abstract: In this paper, the effect of rare earth Ce content on the morphology, composition, type and size distribution of inclusions in W350 non-oriented silicon steel was investigated by means of ICP-MS (inductively coupled plasma mass spectrometry), SEM/EDS (scanning electron microscope-energy Dispersive Spectrometer), and ASPEX (automated SEM/EDS inclusion analysis). The results showed that with the increase of Ce content in the steel, the modification sequence of inclusions was $CeAlO_3 \rightarrow Ce_2O_2S \rightarrow Ce_xS_y$. The type and size distribution of inclusions in the steel obviously changed with the difference in added Ce content. When the added Ce content in the steel was 10 ppm, 14 ppm, 20 ppm and 30 ppm respectively, the rare earth inclusions were mainly $CeAlO_3$-Ce_2O_2S. Furthermore, when the added Ce content increased to 60 ppm, the rare earth inclusions were mainly Ce_2O_2S with a small amount of $CeAlO_3$ contained in part inclusions. When the added Ce content increased continually to 95 ppm, the rare earth inclusions were mainly Ce_xS_y-Ce_2O_2S. The critical Ce content for the conversion between $CeAlO_3$ and Ce_2O_2S was 41 ppm. To ensure that inclusions transform from $CeAlO_3$ to Ce_2O_2S, the Ce content in the steel should be greater than 41 ppm. Under the current experimental conditions, it was found that when the Ce content was 20 ppm, the number density and proportion of inclusions in the steel were lower, and their average size was larger. When the added Ce content increased to 95 ppm, the number density of inclusions in the steel significantly increased, which deteriorated the steel cleanliness.

Keywords: rare earth Ce; inclusions; modification; non-oriented silicon steel; rare earth oxysulfides

Citation: Wang, H.; Niu, Y.; Ling, H.; Qiao, J.; Zhang, Y.; Zhong, W.; Qiu, S. Modification of Rare Earth Ce on Inclusions in W350 Non-Oriented Silicon Steel. *Metals* **2023**, *13*, 453. https://doi.org/10.3390/met13030453

Academic Editors: Andrii Kostryzhev and Maciej Motyka

Received: 27 December 2022
Revised: 15 February 2023
Accepted: 20 February 2023
Published: 22 February 2023

1. Introduction

High-grade, non-oriented silicon steel, as the functional material for high-end power equipment [1], is widely applied to the core of large generators and high-efficiency, energy-saving motors and high-efficiency, energy-saving appliances and electric vehicle manufacturing because of its magnetic characteristics of low iron loss and high magnetic induction [2–4]. With the implementation and promotion of frequency conversion technology in the home appliance industry, the rise of the new-energy automobile industry and the development of energy-saving, high-efficiency motors, there is an increasing demand for the magnetic properties of high-grade, non-oriented silicon steel with low iron loss and high magnetic induction [5]. The factors affecting the magnetic properties of high-grade, non-oriented silicon steel are mainly manifested in two aspects: one is the precise control of the composition and cleanliness of the molten steel that is also the basis for determining the magnetic properties of high-grade, non-oriented silicon steel [6]; and the other is a suitable plastic processing and heat treatment process, as there are many micro-sized inclusions in steel which can have adverse effect on the magnetic domain movement, favorable texture formation and recrystallization process in silicon steel [7]. Rare earth elements have attracted the attention of metallurgical workers much more, because of their unique metallurgical properties. By adding a small amount of rare earth elements into molten steel, the active oxygen and sulfur in the molten steel can be reduced to a lower level, and the microstructure of the finished product can be improved by refining the solidification structure. Meanwhile, rare earth elements can also change the inherent type and morphology of inclusions, reduce the harm of inclusions in experimental molten steel, and improve the related properties of steel products [8–14]. Takashima et al. [15,16] proposed the addition of composite rare earth and Al to non-oriented silicon steel. The results showed that after adding rare earth alloy and Al, the inclusion size became larger, the grain growth rate was significantly improved, the residual stress after annealing was reduced, and the product performance was significantly improved. Wu [17], Liu [18], and Yuan [19] have investigated the application of rare earth elements in non-oriented silicon steel. The research content was mainly based on the addition of pure rare earth elements or rare earth alloy-modified inclusions. Due to their strong deoxidation and desulfurization ability, it was easy to generate rare earth oxides, rare earth sulfides and rare earth oxysulfide compounds by adding appropriate amount of rare earth elements after pre-deoxidation of molten steel. By adding rare earth elements or trace amounts of alloying elements, this could increase the size of inclusions, weaken the pinning effect of original fine inclusions on grain boundaries and magnetic domain walls, make the magnetization process easier, reduce hysteresis loss, and achieve the combination of high magnetic induction and low iron loss. The dispersion precipitation of rare earth elements in molten steel can refine the solidification structure and increase the beneficial texture composition, further improving the magnetic properties of the product. However, it is still unclear whether the mechanism of high melting point phases, such as rare earth oxides (sulfides) as nucleation particles to modify inclusions, can change the type, size and distribution of inclusions in non-oriented silicon steel. Therefore, in this work, efforts have been made to investigate the modification of rare earth Ce on inclusions in W350 non-oriented silicon steel in the laboratory with thermodynamic calculation and experimental analysis, aiming to lay a theoretical foundation for rare earth Ce to modify inclusions in non-oriented silicon steel and provide guidance for production.

2. Materials and Methods

The experimental slab of W350 non-oriented silicon steel was produced by a domestic steel company. The slag was removed cleanly before the hot metal entered the desulfurization station, and the slag was removed twice after the deep desulfurization treatment to $[S] \leq 0.0010\%$, aiming to minimize the high-sulfur slag entering the converter. The scrap steel required self-produced, high-quality scrap steel, of which about 50% was silicon steel scrap. When the converter was tapping, the slide plate and slag stopper were used to stop the slag, and the thickness of the top slag of the ladle was required to be less than 60 mm. Lime was added to the top slag treatment during the tapping process, and the argon flow rate was controlled during the tapping process to prevent the top slag of the ladle from agglomeration. In the RH refining process, it was forbidden to heat up the molten steel by adding aluminum, after the molten steel had arrived at the RH refining station. After the decarburization, the oxygen content of the molten steel was less than 300 ppm before alloying. Following this, the molten steel was cast into a slab. The raw materials used in the experiment were from the continuous casting slab. The chemical composition of raw materials is listed in Table 1.

Table 1. Chemical composition of experimental raw materials (wt.%).

Element	C	Si	Mn	P	S	Als	N
Content	0.0020	2.70	0.32	0.015	0.002	0.50	0.0015

The 150 kg vacuum induction furnace was used for smelting, and the furnace lining was rebuilt before the experiment. The first furnace used pure iron as the raw material to wash the furnace. During the experiment, it was necessary to control the vacuum degree in the furnace, keep the vacuum degree stable; the composition was then adjusted before tapping and rare earth cerium (purity 99.99%) was added, as shown in Table 2. These rare earth cerium alloys were purchased from Beijing Dream Material Technology Co., Ltd. (Beijing, China).

Table 2. Chemical composition of rare earth Ce alloy (wt.%).

Element	Fe	Mg	Ni	Si	Ca	C	W	Ce/RE
Content	<0.050	<0.050	<0.050	0.012	0.023	<0.010	0.035	99.9

The steel composition was detected using inductively coupled plasma mass spectrometry (ICP-MS). Table 3 shows the chemical composition of the experimental steel. The rare earth Ce content in 1#~6# experimental steels was 10 ppm, 14 ppm, 20 ppm, 30 ppm, 60 ppm, and 95 ppm, respectively. Samples (20 mm × 15 mm × 15 mm) were taken from the edge of the slab after removing the oxide scale of the slab surface, as shown in Figure 1. After grinding and polishing, the morphology, size and composition of inclusions in the steel were detected and analyzed using automated SEM/EDS inclusion analysis (ASPEX), scanning electron microscope (SEM) and energy dispersive spectrum (EDS) apparatus. The content of total oxygen (T.O.) and nitrogen in the steel were detected by inert gas fusion pulse-infrared absorption spectroscopy.

Table 3. Chemical composition of experimental steel (wt.%).

Specimen	C	Si	Mn	P	S	Als	N	O	Ce
1#	0.006	2.65	0.17	0.013	0.002	0.22	0.0017	0.0005	0.0010
2#	0.008	2.30	0.22	0.003	0.002	0.58	0.0018	0.0007	0.0014
3#	0.005	2.67	0.19	0.013	0.002	0.44	0.0020	0.0006	0.0020
4#	0.012	2.30	0.22	0.003	0.002	0.58	0.0020	0.0008	0.0030
5#	0.008	2.71	0.18	0.013	0.002	0.41	0.0025	0.0004	0.0060
6#	0.002	2.31	0.23	0.004	0.002	0.49	0.0025	0.0004	0.0095

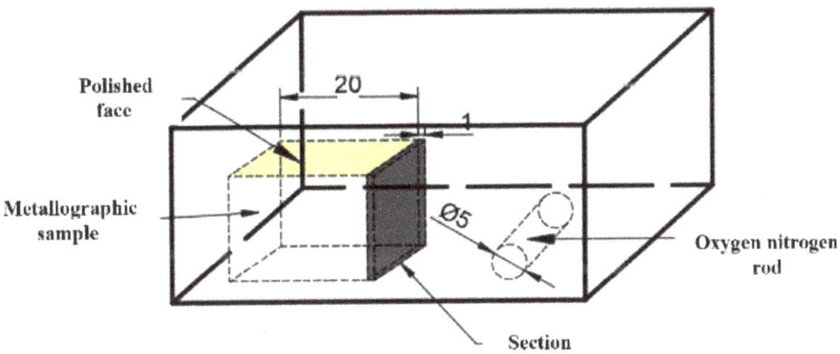

Figure 1. Sampling diagram of the experimental steel.

3. Results

Figure 2 shows the typical morphology and energy spectrum of inclusions in the steel with 10 ppm Ce content. It can be seen that the type of inclusions in the steel was mainly $CeAlO_3$-Ce_2O_2S-(AlN), where $CeAlO_3$-Ce_2O_2S are light gray, and AlN in some inclusions was coated with $CeAlO_3$-Ce_2O_2S. The main morphology of $CeAlO_3$-Ce_2O_2S-(AlN) inclusions was spherical or ellipsoidal. The elemental mapping of a typical inclusion is shown in Figure 3.

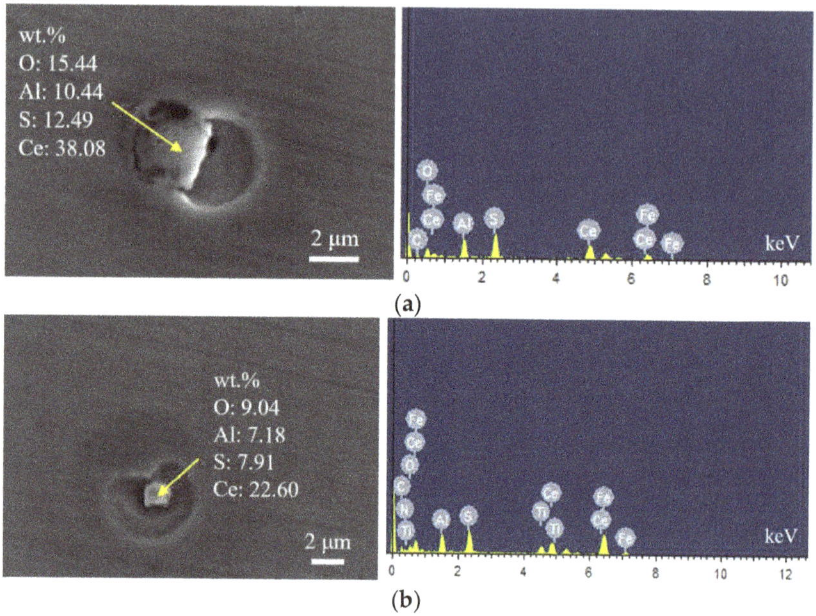

Figure 2. Typical morphology and energy spectrum of inclusions in the steel with 10 ppm Ce content. (**a**) $CeAlO_3$-Ce_2O_2S, (**b**) $CeAlO_3$-Ce_2O_2S.

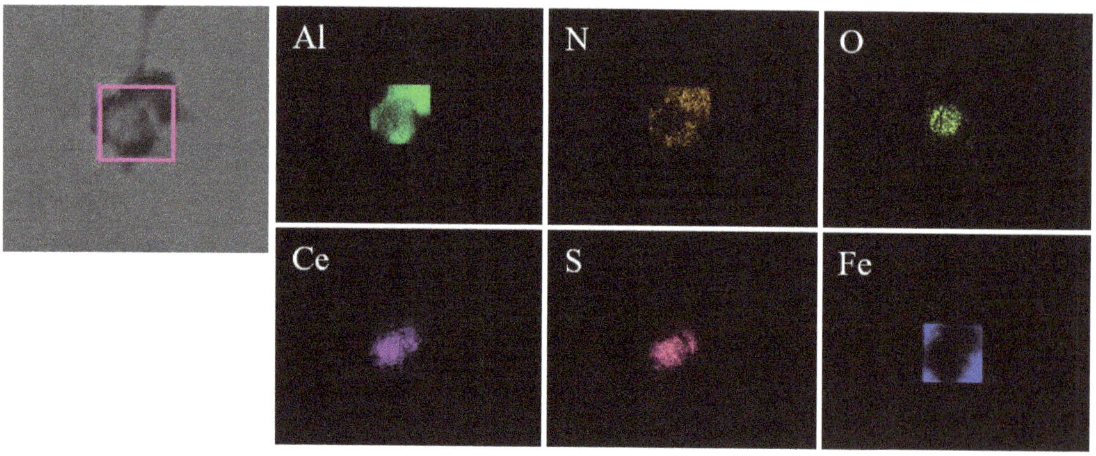

Figure 3. Elemental mapping of a typical inclusion in the steel with 10 ppm Ce content.

Figure 4 presents the typical morphology and energy spectrum results of inclusions in the steel with 14 ppm Ce content. The inclusions were composite phases, mainly with $CeAlO_3$-Ce_2O_2S as the core, and the outer layer was wrapped with AlN. The core of some inclusions was Ce_xS_y-$CeAlO_3$ and its morphology was spherical. The outer layer was wrapped with sharp-angled Al_2O_3-SiO_2. The elemental mapping of a typical inclusion is shown in Figure 5.

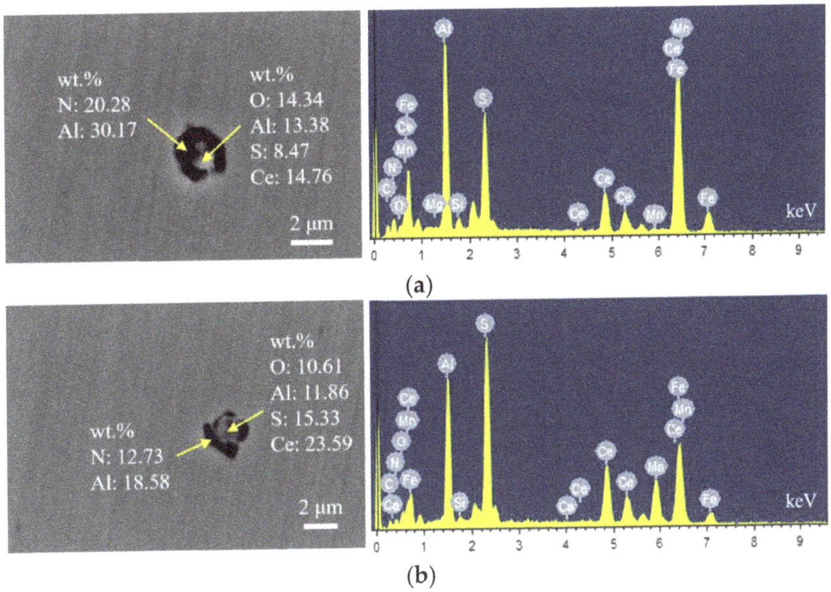

Figure 4. Typical morphology and energy spectrum of inclusions in the steel with 14 ppm Ce content. (**a**) Inner layer: $CeAlO_3$-Ce_2O_2S, Outer layer: AlN, (**b**) Inner layer: $CeAlO_3$-Ce_2O_2S, Outer layer: AlN.

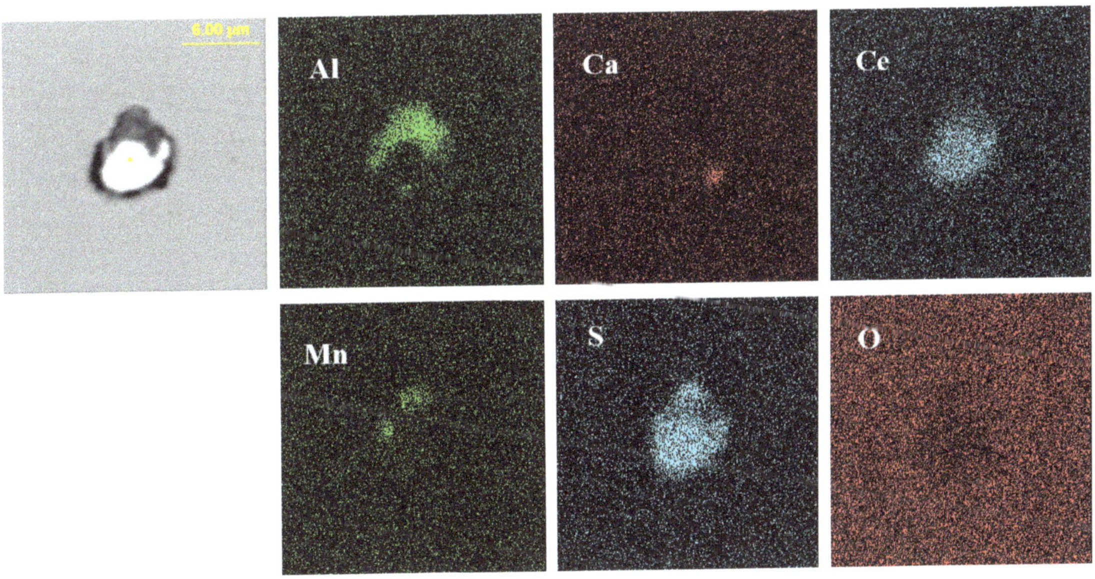

Figure 5. Elemental mapping of a typical inclusion in the steel with 14 ppm Ce content.

Figure 6 shows the typical morphology and energy spectrum results of inclusions in the steel with 20 ppm Ce content. The inclusions in the steel were mainly composite phases, with $CeAlO_3$-Ce_2O_2S as the core. The morphology of inclusions was spherical or ellipsoidal, and the outer layer was wrapped with AlN. The elemental mapping of a typical inclusion is shown in Figure 7.

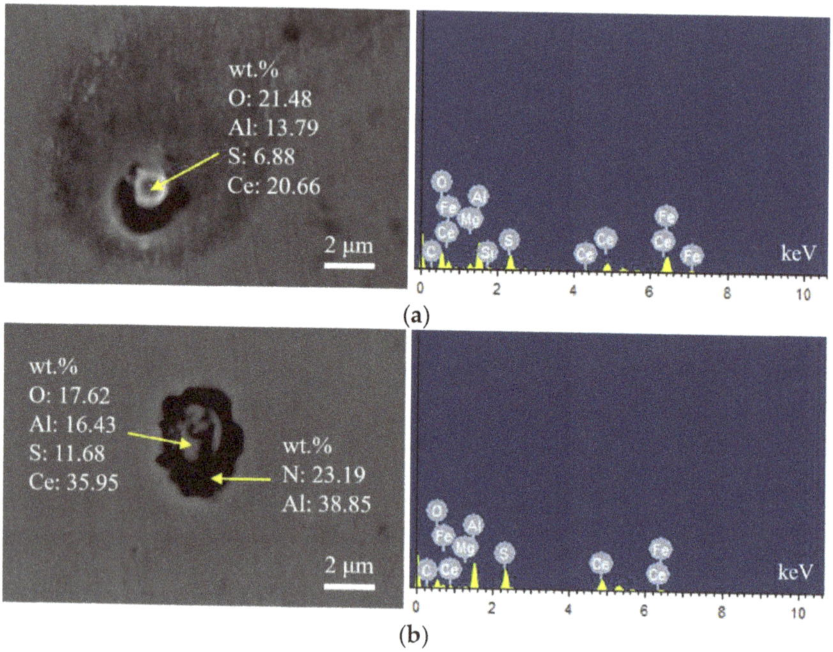

Figure 6. Typical morphology and energy spectrum of inclusions in the steel with 20 ppm Ce content. (**a**) $CeAlO_3$-Ce_2O_2S, (**b**) Inner layer: $CeAlO_3$-Ce_2O_2S, Outer layer: AlN.

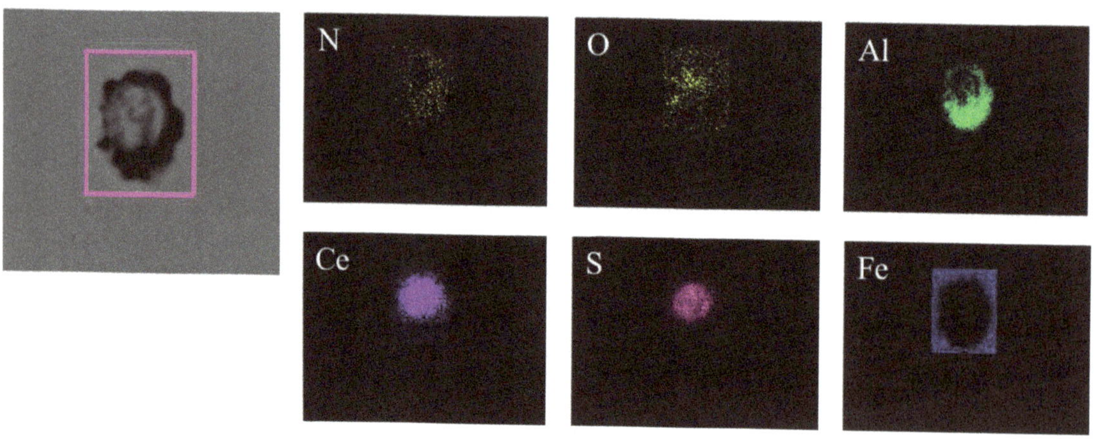

Figure 7. Elemental mapping of a typical inclusion in the steel with 20 ppm Ce content.

Figure 8 shows the typical morphology and energy spectrum results of inclusions in the steel with 30 ppm Ce content. It can be seen that the main types of inclusions in the steel were $CeAlO_3$-Ce_2O_2S-AlN. The inner layer was wrapped with $CeAlO_3$-Ce_2O_2S and its morphology was spherical. The outer layer was wrapped with irregularly shaped AlN. The elemental mapping of a typical inclusion is shown in Figure 9.

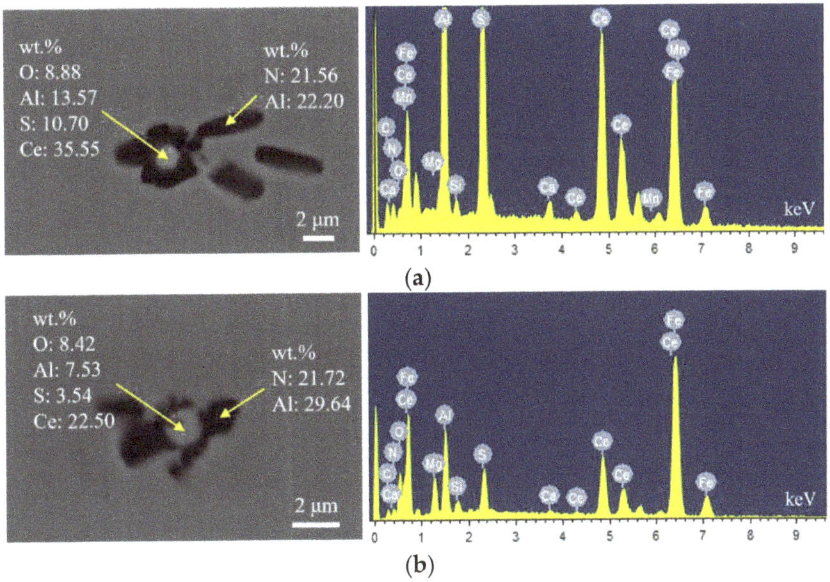

Figure 8. Typical morphology and energy spectrum of inclusions in the steel with 30 ppm Ce content. (**a**) Inner layer: $CeAlO_3$-Ce_2O_2S, Outer layer: AlN, (**b**) Inner layer: $CeAlO_3$-Ce_2O_2S, Outer layer: AlN.

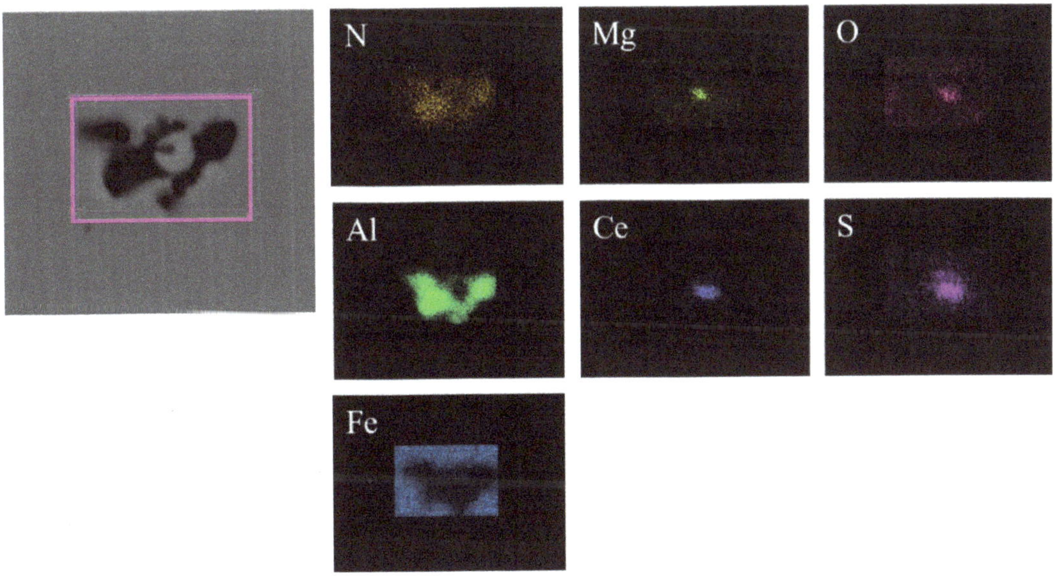

Figure 9. Elemental mapping of a typical inclusion in the steel with 30 ppm Ce content.

Figure 10 shows the typical morphology and energy spectrum results of inclusions in the steel with 60 ppm Ce content. It was found that when the Ce content increased to 60 ppm, the types of inclusions in the steel were mainly $CeAlO_3$-Ce_2O_2S-AlN and Ce_2O_2S-AlN. The morphology of Ce_2O_2S inclusions was approximately spherical. As shown in Figure 11, the core of the inclusion was Ce_xS_y-Ce_2O_2S and the outer layer was wrapped with $MgO{\cdot}Al_2O_3$ phase.

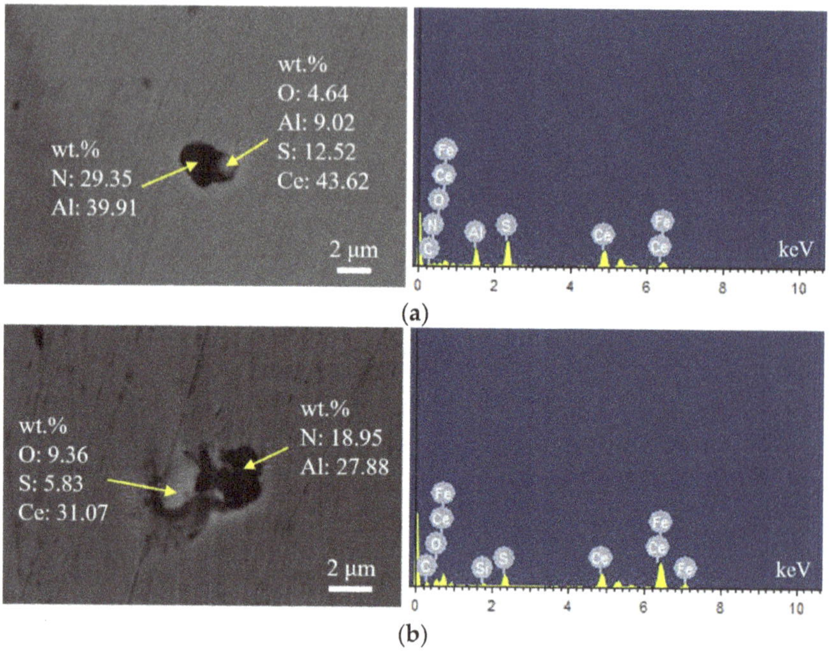

Figure 10. Typical morphology and energy spectrum of inclusions in the steel with 60 ppm Ce content. (**a**) $CeAlO_3$-Ce_2O_2S-AlN, (**b**) Ce_2O_2S-AlN.

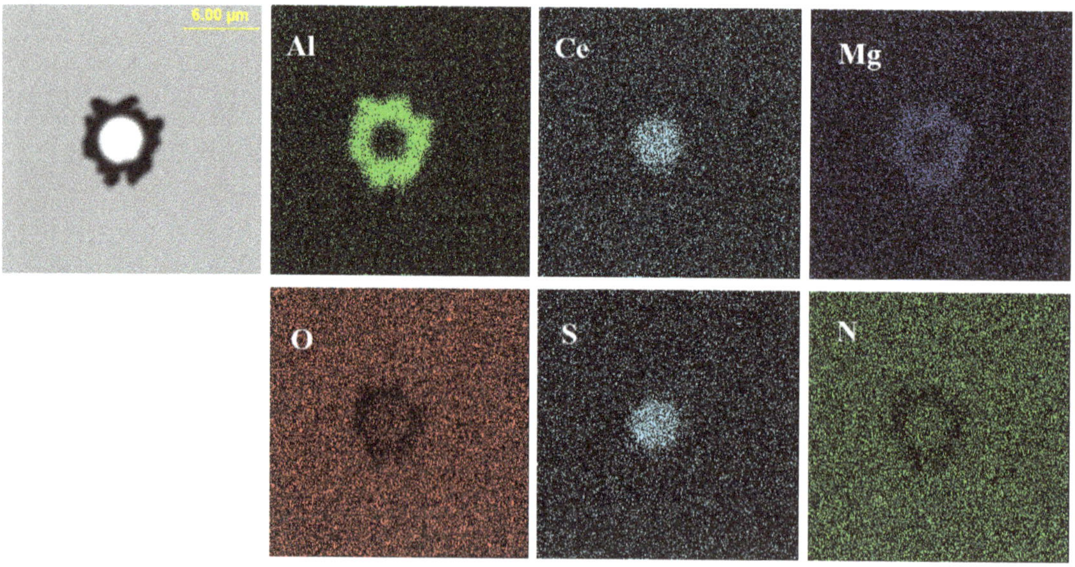

Figure 11. Elemental mapping of a typical inclusion in the steel with 60 ppm Ce content.

Figure 12 presents the typical morphology and composition of inclusions in the steel with 95 ppm Ce content. When the Ce content increased to 95 ppm, the core of inclusions in the steel was Ce_xS_y-Ce_2O_2S and its morphology was spherical or ellipsoidal. The outer layer of some inclusions was wrapped with AlN or $MgO{\cdot}Al_2O_3$ phase. The elemental mapping of a typical inclusion is shown in Figure 13.

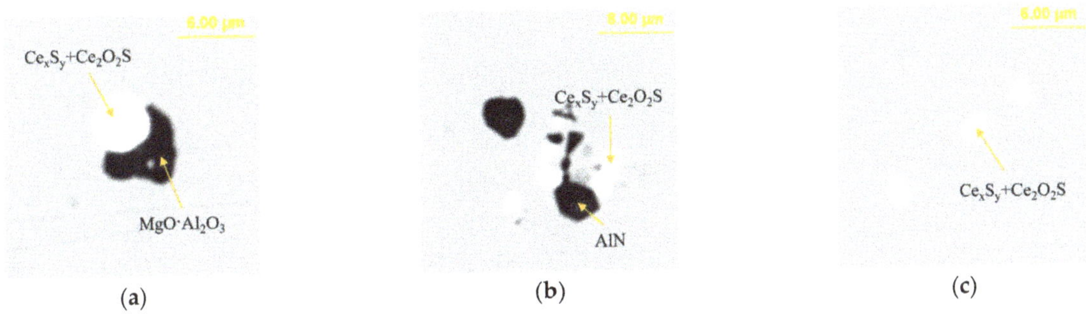

Figure 12. Typical morphology and energy spectrum of inclusions in the steel with 95 ppm Ce content. (**a**) Ce_xS_y-Ce_2O_2S-$MgO{\cdot}Al_2O_3$, (**b**) Ce_xS_y-Ce_2O_2S-AlN, (**c**) Ce_xS_y-Ce_2O_2S.

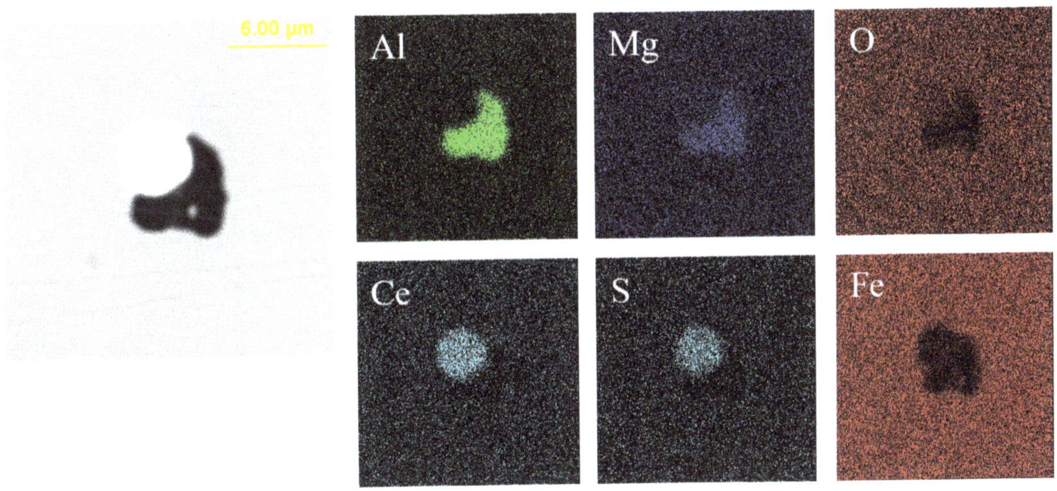

Figure 13. Elemental mapping of a typical inclusion in the steel with 95 ppm Ce content.

4. Discussion

4.1. Effects of Rare Earth Ce Content on Inclusion Type

The type of inclusions in W350 non-oriented silicon steel was mainly Al_2O_3 without rare earth Ce. With the addition of Ce, reactions with O and S in molten steel would occur to form a variety of oxides, sulfides and oxysulfides. The possible reactions involved are as follows [20–24]:

$$2[\mathrm{Al}] + 3[\mathrm{O}] = (\mathrm{Al_2O_3}) \quad \Delta G^{\theta} = -1202000 + 386.3T\ \mathrm{J/mol} \tag{1}$$

$$[\mathrm{Ce}] + 2[\mathrm{O}] = (\mathrm{CeO_2}) \quad \Delta G^{\theta} = -852720 + 249.96T\ \mathrm{J/mol} \tag{2}$$

$$[\mathrm{Ce}] + \frac{3}{2}[\mathrm{O}] = \frac{1}{2}(\mathrm{Ce_2O_3}) \quad \Delta G^{\theta} = -714380 + 179.74T\ \mathrm{J/mol} \tag{3}$$

$$[\mathrm{Ce}] + [\mathrm{S}] = (\mathrm{CeS})\ \Delta G^{\theta} = -422100 + 120.38T\ \mathrm{J/mol} \tag{4}$$

$$[\mathrm{Ce}] + \frac{3}{2}[\mathrm{S}] = \frac{1}{2}(\mathrm{Ce_2S_3}) \quad \Delta G^{\theta} = -536420 + 163.86T\ \mathrm{J/mol} \tag{5}$$

$$[\mathrm{Ce}] + \frac{4}{3}[\mathrm{S}] = \frac{1}{3}(\mathrm{Ce_3S_4}) \quad \Delta G^{\theta} = -497670 + 146.30T\ \mathrm{J/mol} \tag{6}$$

$$[\mathrm{Ce}] + [\mathrm{O}] + \frac{1}{2}[\mathrm{S}] = \frac{1}{2}(\mathrm{Ce_2O_2S}) \quad \Delta G^{\theta} = -675700 + 165.5T\ \mathrm{J/mol} \tag{7}$$

$$[Ce] + [Al] + 3[O] = (CeAlO_3)\ \Delta G^{\theta} = -1366460 + 364.3T\ J/mol \quad (8)$$

$$[Ce] + (Al_2O_3) = (CeAlO_3) + [Al]\ \Delta G^{\theta} = -162837 - 22.4T\ J/mol \quad (9)$$

$$[Ce] + [S] + (CeAlO_3) = (Ce_2O_2S) + [Al] + [O]\ \Delta G^{\theta} = 12870 - 32.3T\ J/mol \quad (10)$$

$$2[Al] + 3[O] + (Ce_2O_3) = 2(CeAlO_3)\ \Delta G^{\theta} = -1301800 + 368.6T\ J/mol \quad (11)$$

$$(Ce_2O_3) + [S] = (Ce_2O_2S) + [O]\ \Delta G^{\theta} = 77530 - 28T\ J/mol \quad (12)$$

When the rare earth Ce is fed into molten steel, it not only reacts easily with Al_2O_3 to form $CeAlO_3$, but also converts the generated Ce-containing inclusions. The possible reactions involved are as follows:

$$2(CeAlO_3) + 2[Al] + 3[S] = 2(Al_2O_3) + (Ce_2S_3)\ \Delta G^{\theta} = -743920 + 371.72T\ J/mol \quad (13)$$

$$3(Ce_2O_2S) + 2[Al] + 3[S] = 2(CeAlO_3) + 2(Ce_2S_3)\ \Delta G^{\theta} = -824400 + 391.04T\ J/mol \quad (14)$$

$$3(Ce_2O_2S) + 4[Al] + 6[S] = 2(Al_2O_3) + 3(Ce_2S_3)\ \Delta G^{\theta} = -1568320 + 762.76T\ J/mol \quad (15)$$

$$2(CeS) + [S] = (Ce_2S_3)\ \Delta G^{\theta} = -228640 + 86.96T\ J/mol \quad (16)$$

The general expression of Gibbs free energy is:

$$\Delta G = \Delta G^{\theta} + RT \ln J \quad (17)$$

where J is the ratio of the activity product of the resultant and the activity product of the reactant.

Assuming that the molten steel is an ideal dilute solution, and the solute follows Henry's law, one can satisfy the following formula for calculating activity and activity coefficient [25]:

$$a_i = f_i \cdot w[i] \quad (18)$$

where a_i is the activity of component i, f_i is the activity coefficient of component i, and $w_{[i]}$ is the mass percentage of the element i. Thus:

$$\lg f_i = \sum_{j=1}^{n} e_i^j \cdot w[j] \quad (19)$$

where e_i^j is the interaction coefficient between j element and component i in molten steel. The first-order interaction coefficient of each element in molten steel is listed in Table 4 [26–28].

Table 4. The first-order interaction coefficient of each element in molten steel.

e_i^j	C	Si	Mn	P	S	Al	O	Ce	N
S	0.11	0.063	−0.026	0.029	−0.028	0.035	−0.27	−0.856	0.01
Al	0.091	0.0056	0.012	0.05	0.03	0.045	−6.6	−0.43	−0.058
O	−0.45	−0.131	−0.021	0.07	−0.133	−3.9	−0.2	−0.57	0.057
Ce	−0.077	-	0.13	1.77	−39.8	−2.25	−5.03	−0.003	-

Figure 14 shows the stability diagram of Al_2O_3-$CeAlO_3$-Ce_2O_2S-Ce_xS_y in non-oriented electrical steel. Obviously, with the increase in Ce content in the steel, the evolution sequence of inclusions was $CeAlO_3 \rightarrow Ce_2O_2S \rightarrow Ce_xS_y$. Based on the experimental results, the types of inclusions in the steel were mainly $CeAlO_3$-Ce_2O_2S-(AlN) when the Ce content was 10 ppm. When the Ce content was 14 ppm, the inclusions were composed of composite phases. The core was mainly $CeAlO_3$-Ce_2O_2S and the outer layer was wrapped with AlN. Meanwhile, the core of some inclusions was Ce_xS_y-$CeAlO_3$. When the Ce content was 20 ppm, the composition of inclusions was also composite phases with $CeAlO_3$-Ce_2O_2S as the core. When the Ce content was increased to 30 ppm, the main types of inclusions in the steel were $CeAlO_3$-Ce_2O_2S-AlN. The above results showed that the rare earth inclusions in the steel were mainly $CeAlO_3$ + Ce_2O_2S when the added Ce content was 10~30 ppm.

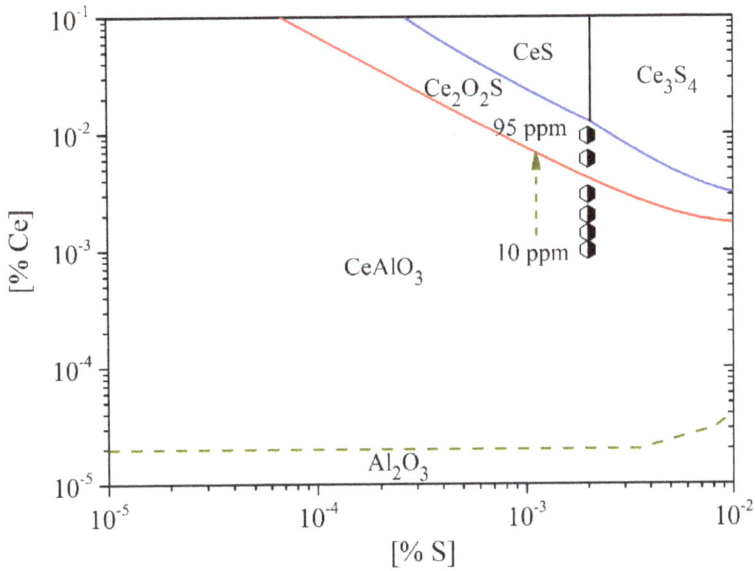

Figure 14. Stability diagram of Al_2O_3-$CeAlO_3$-Ce_2O_2S-Ce_xS_y in non-oriented electrical steel.

Furthermore, when the Ce content was increased to 60 ppm, the composition of inclusions in the steel was mainly Ce_2O_2S and some inclusions contained a small amount of $CeAlO_3$. The inclusions were changed to Ce_xS_y-Ce_2O_2S when the Ce content was 95 ppm. As shown in Figure 14, when the added Ce content is 95 ppm, the composition of the inclusions is located near the boundary between Ce_xS_y and Ce_2O_2S, which was consistent with the experimental results.

Figure 14 also shows that the critical Ce content for the conversion between $CeAlO_3$ and Ce_2O_2S was 41 ppm. To ensure the inclusions transformed from $CeAlO_3$ to Ce_2O_2S, the Ce content in the steel had to be greater than 41 ppm. Due to the addition of rare earth Ce elements, the formation of rare earth aluminates, rare earth oxygen sulfides and rare earth sulfides with high melting point can occur in molten steel. These large-sized composite inclusions can be used as heterogeneous nucleation particles and it is beneficial to control the modification and precipitation of Al_2O_3, MnS, TiN, and AlN, thereby improving the magnetic properties of the product.

4.2. Effects of Rare Earth Ce on Size and Distribution of Inclusions

Figure 15 shows the size distribution and the average size of inclusions in steel of different heats. It shows that when the added Ce content in the steel was 10 ppm, the size of inclusions in the steel was concentrated in the range of 2.0~4.5 μm and the corresponding proportion was 10.3~21.4%. When the added Ce content was 20 ppm and 60 ppm, the number density of inclusions in the steel was relatively low (11.11 #/mm^2 and 15.14 #/mm^2, respectively). Addition of Ce decreased the amount of harmful inclusions (such as Al_2O_3, MnS) in the steel, which is beneficial for improving the steel cleanliness and the magnetic properties of the product. Moreover, the number density of >5.0 μm inclusions significantly increased (6.9 #/mm^2 and 6.5 #/mm^2 with Ce content of 20 ppm and 60 ppm, respectively), indicating that rare earth treatment can promote the generation of large-sized inclusions in the steel. When the Ce content was 20 ppm, the average size of inclusions in the steel was at its largest, at 6.9 μm. When the Ce content in the steel increased to 95 ppm, the number density of inclusions of all size ranges obviously increased and the total number density was 130.1 #/mm^2. It was larger than that of other Ce content levels, especially for inclusions with size range of 1.0~1.5 μm and >5.0 μm. Their number densities were 33.17 #/mm^2 and 33.54 #/mm^2, respectively, and the corresponding proportion was about 25%, which would deteriorate the steel cleanliness and the magnetic properties of the product.

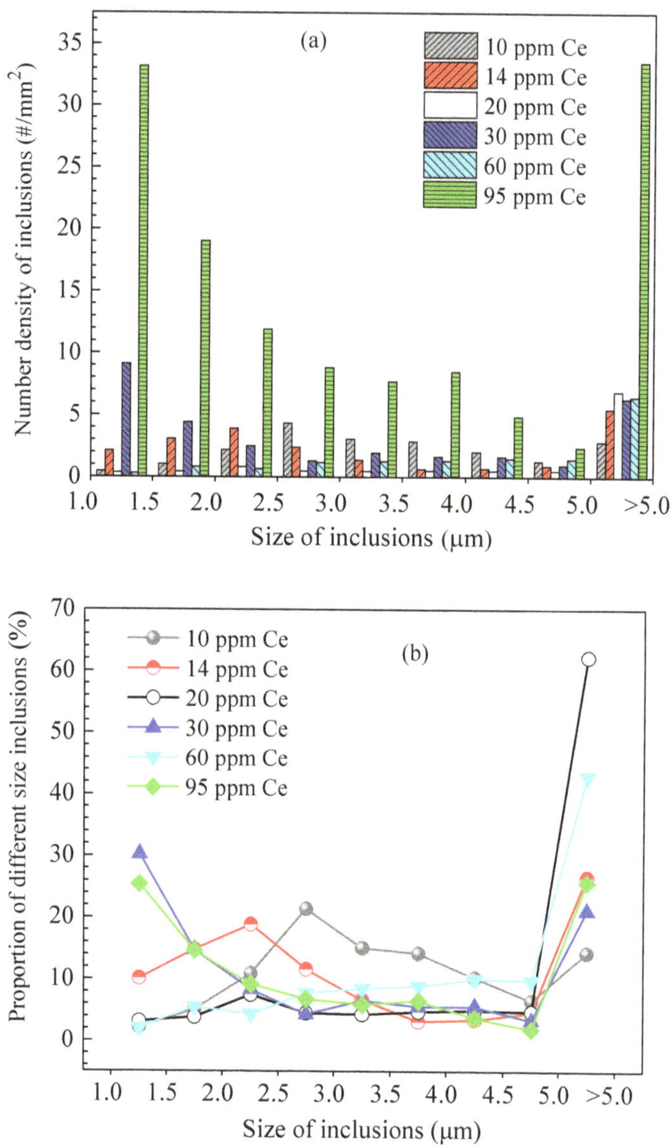

Figure 15. *Cont.*

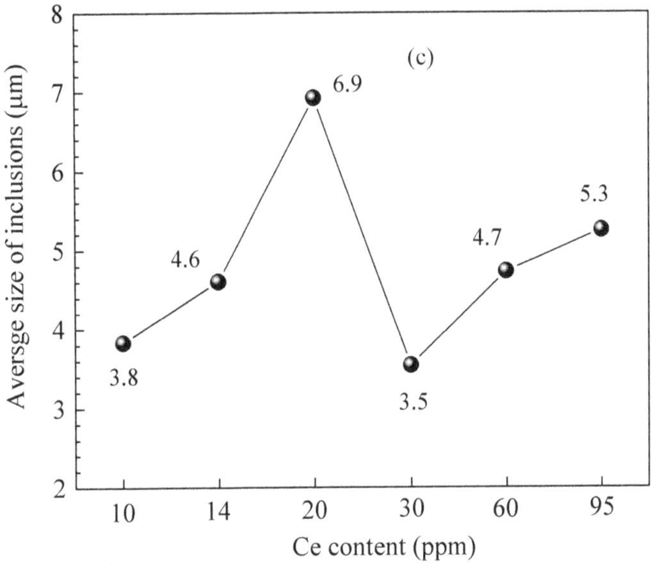

Figure 15. Size distribution and average size of inclusions in the steel of different heats. (**a**) Inclusion size distribution, (**b**) ratio of different size inclusions, (**c**) average size of inclusions.

5. Conclusions

(1) With the increase in the Ce content in the steel, the modification sequence of inclusions was $CeAlO_3 \rightarrow Ce_2O_2S \rightarrow Ce_xS_y$. When the added Ce content in the steel was 10 ppm, 14 ppm, 20 ppm and 30 ppm respectively, the rare earth inclusions were mainly $CeAlO_3$-Ce_2O_2S. When the added Ce content increased to 60 ppm, the rare earth inclusions were mainly Ce_2O_2S and a small amount of $CeAlO_3$ contained in part inclusions. When the added Ce content increased continually to 95 ppm, the rare earth inclusions were mainly Ce_xS_y + Ce_2O_2S.

(2) When the added Ce content was 95 ppm, the composition was located near the boundary between rare earth sulfide and rare earth oxysulfide. The critical Ce content for the conversion between $CeAlO_3$ and Ce_2O_2S was 41 ppm.

(3) The difference in the Ce content significantly affected the number density and size distribution of inclusions in the steel. When the added Ce content was 20 ppm, the number density and proportion of inclusions in the steel were lower, and its average size was larger. When the added Ce content increased to 95 ppm, the number density of inclusions in the steel significantly increased.

Author Contributions: Conceptualization, H.W. and H.L.; methodology, H.W.; software, H.W.; validation, H.W., Y.Z. and W.Z.; formal analysis, H.W.; investigation, H.W.; resources, H.W.; data curation, H.W., J.Q. and Y.N.; writing—original draft preparation, H.W.; writing—review and editing, H.W. and S.Q.; visualization, H.W. and H.L.; supervision, S.Q.; project administration, H.W. and S.Q.; funding acquisition, H.W. All authors have read and agreed to the published version of the manuscript.

Funding: This research was funded by the National Natural Science Foundation of China (Nos. 52274312 and 51804003), the Fund of Education Department of Anhui Province (Nos. 2022AH050291, 2022AH050293), the Open Project Program of Anhui Province Key Laboratory of Metallurgical Engineering & Resources Recycling (Anhui University of Technology) (No. SKF21-04), and the Jiangxi Province Major Scientific and Technological Research and Development Special Funding Project (20213AAE01009).

Institutional Review Board Statement: Not applicable.

Informed Consent Statement: Not applicable.

Data Availability Statement: Data available in a publicly accessible repository.

Conflicts of Interest: The authors declare no conflict of interest.

References

1. Ji, S.C.; Shi, Y.H.; Zhang, F.; Lu, W.F. Review on vibration and noise of power transformer and its control measures. *High Volt. Appar.* **2019**, *55*, 1–17.
2. Yan, M.; Xiao, H.H.; Huang, X.; Zhang, Y.D. Analysis of new energy vehicles development and electrical steel demand for driving motor. *Electr. Steel* **2020**, *2*, 1–5.
3. Xu, X.X.; Qin, J.; Zhao, H.B.; Nie, J.C.; Yu, H.Y. A study on the research status and development trend of non-oriented electrical steel with high grade for new energy vehicles. *Jiangxi Metall.* **2020**, *40*, 6–11.
4. Gao, L.Y.; Zhou, H.J.; Wang, C.; Pei, R.L. Development overview and trend analysis of electric drive system of new energy vehicle. *Electr. Steel* **2020**, *2*, 27–31.
5. Chen, Z. Research on the development change and the demand of electrical steel in China in the new era. *Electr. Steel* **2019**, *1*, 1–6.
6. Guo, F.H. Study on Control of Oxide Inclusion in Non-Oriented Electrical Steel. Master's Thesis, Anhui University of Technology, Ma'anshan, China, 3 June 2019.
7. He, Z.Z.; Zhao, Y.; Luo, H.W. *Electrical Steel*; Metallurgical Industry Press: Beijing, China, 2012.
8. Zhang, X.F.; Tang, J.P.; Han, C.P.; Wang, A.L.; Zhi, J.G. Analysis on the role of rare earth in steel and the present situation of industrial production. *Chin. Rare Earths* **2020**, *42*, 117–130.
9. Dong, H.; Lian, X.T.; Hu, C.D.; Lu, H.C.; Peng, W.; Zhao, H.S.; Xu, D.X. High performance steels: The scenario of theory and technology. *Acta Metall. Sin.* **2020**, *56*, 558–582.
10. Ren, H.P.; Wu, Z.W.; Jin, Z.L. Functions and research progresses of rare earth in electrical steel. *Sci. Technol. Baotou Steel* **2018**, *44*, 1–6.
11. Wang, L.M.; Tan, Q.Y.; Li, N.; Qin, Z.; Qiu, S.T.; Gan, Y. Research progress on rare earth application technology in non-oriented electrical steels. *J. Chin. Soc. Rare Earths* **2014**, *32*, 513–533.
12. Zhang, F.; Li, G.Q.; Zhu, C.Y. Heredity characters of inclusions and electromagnetic properties of non-oriented silicon steel treated by rare earth alloy. *J. Funct. Mater.* **2013**, *44*, 1029–1033.
13. Lin, Q.; Guo, F.; Zhu, X.Y. Behaviors of La and Ce on grain boundaries in carbon manganese cleanliness steel. *J. Chin. Soc. Rare Earths* **2006**, *24*, 427–430.
14. Wang, L.M.; Du, T.; Lu, X.L.; Yue, K.X. Study of behaviors and application of micro-rare earth elements in steel. *Chin. Rare Earths* **2001**, *22*, 37–40.
15. Takashima, M.; Morito, N.; Honda, A.; Maeda, C. Non-oriented electrical steel sheet with low iron loss for high-efficiency motor cores. *IEEE Trans. Magn.* **1999**, *35*, 557–561. [CrossRef]
16. Takashima, M.; Ono, T.; Nishimura, K. Low iron loss non-oriented electrical steel for high efficiency motors "RMA Series". *Kawasaki Steel Tech. Rep.* **1998**, *39*, 48–49.
17. Wu, C.J.; Tang, X.L. Grain boundary segregation of Ce, Mo and P and temper embrittlement of steel. *Iron Steel* **1991**, *26*, 31–35.
18. Liu, H.M.; Gao, L. The effect of rare earth metals in non-oriented silicon steel. *Iron Steel* **1987**, *22*, 41–45.
19. Yuan, Z.X.; Li, J.H.; Feng, S.J.; Wu, C.J. Grain boundary segregation of Ce and temper embrittlement of steel. *Chin. J. Eng.* **1982**, *S1*, 42–50.
20. Chen, J.X. *Manual of Chart Data for Steelmaking*; Metallurgical Industry Press: Beijing, China, 2010.
21. Itoh, H.; Hino, M.; Ban-Ya, S. Thermodynamics on the formation of spinel nonmetallic inclusion in liquid steel. *Metall. Mater. Trans. B* **1997**, *28*, 953–956. [CrossRef]
22. Vahed, A.; Kay, D.A.R. Thermodynamics of rare earths in steelmaking. *Metall. Mater. Trans. B* **1976**, *7*, 375–383. [CrossRef]
23. Li, W.C. Thermodynamics of the formation of rare-earth inclusions in steel. *Iron Steel* **1986**, *21*, 7–12.
24. Wang, L.M.; Du, T.; Lu, X.L.; Li, Z.B.; Gai, Y.C. Thermodynamics and application of rare earth elements in steel. *J. Chin. Soc. Rare Earths* **2003**, *21*, 251–254.
25. Huang, X.H. *Principles of Iron and Steel Metallurgy*; Metallurgical Industry Press: Beijing, China, 2013.
26. Zhang, L.F.; Ren, Y.; Duan, H.J.; Yang, W.; Sun, L.Y. Stability diagram of Mg-Al-O system inclusions in molten steel. *Metall. Mater. Trans. B* **2015**, *46*, 1809–1825. [CrossRef]
27. Wang, Y.; Li, C.R.; Zeng, Z.Y.; Xi, Z.B. Crystallization of alumina modified by rare earth elements in SWRS82B steel. *Iron Steel* **2020**, *55*, 69–74.
28. Pan, N.; Song, B.; Zhai, Q.J.; Wen, B. Effect of lattice disregistry on the heterogeneous nucleation catalysis of liquid steel. *J. Beijing Univ. Sci. Technol.* **2010**, *32*, 179–182.

metals

MDPI

Article

Formation Mechanism and Improvement of Magnetic Particle Inspection Defects in Cr5 Backup Roller Forged Ingot

Weifeng Zhang [1,2], Guanbo Wang [1], Yanling Zhang [1,*], Guoguang Cheng [1] and Zhonghua Zhan [2]

[1] State Key Laboratory of Advanced Metallurgy, University of Science and Technology Beijing, Beijing 100083, China; zhangwf@citic-hic.com.cn (W.Z.); owenbread@xs.ustb.edu.cn (G.W.); chengguoguang@metall.ustb.edu.cn (G.C.)
[2] Luoyang CITIC HIC Casting and Forging Co., Ltd., Luoyang 471039, China; lyzhan1005@xs.ustb.edu.cn
* Correspondence: zhangyanling@metall.ustb.edu.cn

Abstract: Industrial tests and thermodynamic calculations were utilized to investigate the source and formation of magnetic particle inspection defects identified on the near-surface of the Cr5 back-up roll forged ingot, which was used in large cold rolling mills. The results showed that the linear aggregating SiO_2-MnO-Al_2O_3 liquid inclusions up to 3 mm led to the flaw detection failure. SiO_2-MnO-Al_2O_3 liquid inclusions were firstly formed in the inductive furnace. Due to its low contact angle, a huge amount of SiO_2-MnO-Al_2O_3 liquid inclusions were inherited into the forged ingot. The formation of SiO_2-MnO-Al_2O_3 liquid inclusions was attributed to the over-oxidation and relatively low aluminum content in the molten steel, as calculated by Factsage 8.1. Controlling the amount of aluminum in molten steel during the smelting process could modify the formation of SiO_2-MnO-Al_2O_3 and CaO-SiO_2-Al_2O_3 liquid oxide into solid Al_2O_3 type inclusions that were easily removed. Besides, the CaO-SiO_2-Al_2O_3 liquid oxide could be transformed from CaO-Al_2O_3 type oxide by the significant loss of aluminum content during the VD process or slag entrapment. Certain content of aluminum in the molten steel could improve the flaw detection caused by the aggregating SiO_2-MnO-Al_2O_3 inclusions effectively.

Keywords: Cr5; back-up roller; forged ingot; large-size inclusions; SiO_2-MnO-Al_2O_3; contact angle; aluminum deoxidation; magnetic particle inspection; defects

Citation: Zhang, W.; Wang, G.; Zhang, Y.; Cheng, G.; Zhan, Z. Formation Mechanism and Improvement of Magnetic Particle Inspection Defects in Cr5 Backup Roller Forged Ingot. *Metals* **2022**, *12*, 295. https://doi.org/10.3390/met12020295

Academic Editor: Alexander Ivanovich Zaitsev

Received: 6 January 2022
Accepted: 5 February 2022
Published: 8 February 2022

1. Introduction

In recent years, rolling mills have been making progress to meet the ever-increasing needs for high-quality steel products, particularly in China. Accordingly, the work roller and back-up roller are required to exhibit superior performance such as exceptional fatigue strength under high load and speed service conditions [1–7].

It is widely recognized that increasing Chromium content could reduce wear loss linearly [8–11]. Therefore, the chromium content of backup rollers has increased from 1.2% up to 3% and 5% in the 1970s [9,12,13]. The appearance of vacuum refining and electro-slag remelting (ESR) technology guaranteed the goal of high carbon with high alloy elements content and achieved the low hydrogen content [13,14]. Adding alloy elements like Mo, V, and W into backup rollers could also promote mechanical properties further as required. Nowadays, backup rollers were classified into Cr3 series and Cr5 series by their chromium content [8,9,15]. Non-destructive testing (NDT) technology, e.g., ultrasonic testing (UT), magnetic particle inspection (MPI), radiographic testing (RT), etc., have been widely used in the industry [16]. MPI was utilized frequently among these methods owing to the demanding requirements of surface and near-surface quality in rollers More and more defects were considered as metallurgical defects, which were caused by inclusions created in deoxidation, reactions between molten steel, slag and refractory, or even slag entrapment [17–24]. Jamil et al. [24] found that alumina and oxide types non-metallic inclusions caused the failure of ultrasonic testing in the sugar mill shaft. Lu et al. [25]

found that large-sized, long-striped MnS inclusions were the major cause of the magnetic particle testing failure in the heavy truck crankshaft. Wang et al. [26] also found that the aggregation of single Al_2O_3 particles and Al_2O_3 clusters existed on the surface of 80 t 20Cr-8Ni stainless ingot, which failed the penetrating testing.

However, the studies aimed at promoting the properties of rollers were mainly around the after-casting manufacturing procedure, e.g., obtaining ideal characteristics of carbides, microstructure by heat treatment, etc. [27–30]. Meanwhile, few studies concerning the evolution and influence of inclusions in the large forged ingots were found [24]. Consequently, combined the appearance of ultrasonic defects on the post-forging rollers with the corresponding specimens during the plant steelmaking process, the origin, evolution, and improvement of defects have been analyzed and clarified. Besides, under the different deoxidation processes, the corresponding evolution and formation mechanism of oxide inclusions aided by thermodynamics calculation and plant trials were discussed as follows.

2. Materials and Methods

2.1. Experimental Procedure

The steelmaking process of 50ton Cr5 type backup roller could be briefly described as "induction furnace (IF) → ladle furnace (LF1) → ladle furnace (LF2) → vacuum degasser (VD) → ladle furnace (LF3) → vacuum casting (VC)", as shown in Figure 1 The return scrap and alloy raw materials were melted in the IF. Then the deoxidation, desulfurization, and composition adjustment took place in LF1, which served as a pre-refining procedure. After the LF2 refining process, the vacuum degasser mainly undertook to degas especially dehydrogenation. Afterward, LF3 was considered to fulfill the final composition adjustment and superheat degree requirements before vacuum casting. To trace the origin of the defects, we implemented the tracking and analysis during the whole metallurgical process. And the original deoxidizer was silicon-aluminum (Si-Al) complex deoxidizer (Si: 65% in mass%, Al: 25% in mass%).

Lollipop steel specimens quenched by water and slag were sampled at different smelting stages i.e., the end of IF smelting, the beginning and end of each LF refining process, the vacuum-breaking of VD, and the beginning of VC, which were named by their chronological order, respectively. The defective parts sampled after forging and fine machining were rechecked and illustrated by a magnetic particle flaw detector, which was processed into 15 mm × 15 mm × 15 mm cubes to achieve the defective location and characteristics by optical microscope, as shown in Figure 2. All the specimens in the smelting process were machined to 10 mm × 10 mm × 10 mm cubes for testing and analysis. Schematic illustrations of the metallurgical process and sampling were shown in Figure 1.

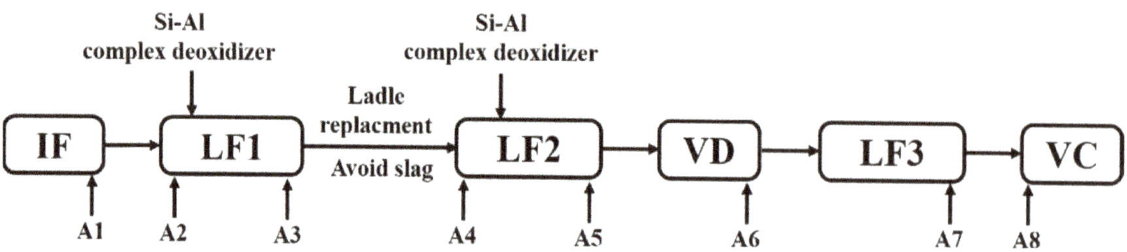

Figure 1. Schematic illustration of original metallurgical process and sampling.

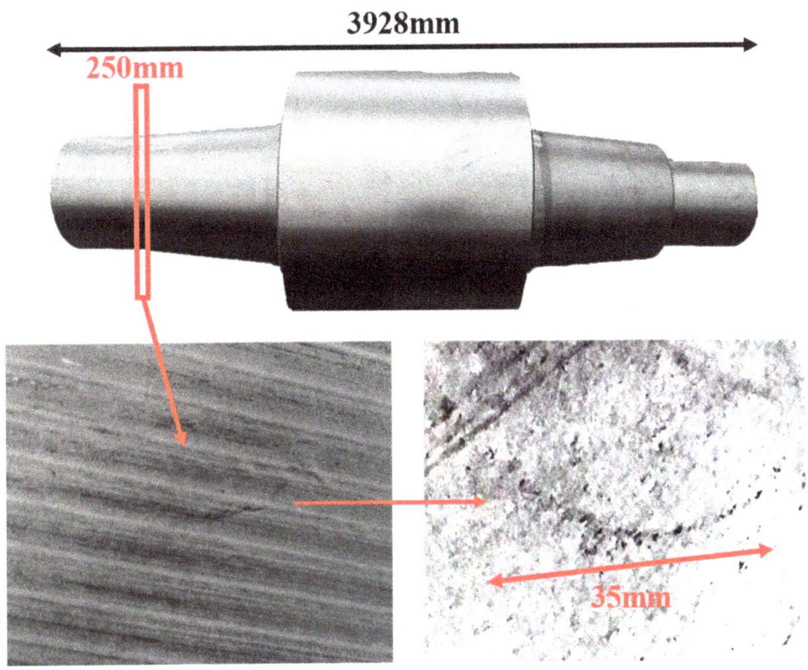

Figure 2. Sampling and defective part characteristics on the roller.

2.2. Analysis Method of Composition and Micro-Inspection

The [Si], [Mn], [Al], [Ca] contents in molten steel were determined by inductively coupled plasma optical emission spectrometer method (ICP-OES). (ThermoFisher, Waltham, MA, USA) The total oxygen contents were analyzed by the inert gas fusion-infrared absorptiometry method with an accuracy of ±0.0001%. The compositions of slag specimens that corresponded with steel specimens, respectively, were determined by X-ray fluorescence spectrometer (XRF). (PANalytical B.V., Almelo, The Netherlands) Additionally, the steel specimens were all well-ground and polished for the following micro-inspection. Scanning electron microscopy (SEM) (FEI, Hillsboro, OR, USA) and energy-dispersive X-ray spectroscopy (EDS) (Bruker, Billerica, MA, USA) for the micro-inspection. The maximum diameter of inclusions was considered as the size of the inclusions. Afterward, the composition of steel and slag specimens is shown in Tables 1 and 2, respectively.

Table 1. Main chemical composition of steel specimens in mass %.

Specimen No.	Stage	C/%	Si/%	Mn/%	P/%	S/%	Cr/%	Al/%	Ca/%	O/%
A1	IF-end	0.407	0.23	0.29	0.0095	0.0034	3.32	0.0018	0.0003	0.0242
A2	LF1-initial	0.422	0.22	0.29	0.0092	0.0034	3.27	0.0015	0.0008	0.0285
A3	LF1-end	0.433	0.19	0.32	0.0098	0.0025	3.30	0.0036	0.0003	0.0045
A4	LF2-initial	0.423	0.17	0.32	0.0098	0.003	3.24	0.0032	0.0003	0.0067
A5	LF2-end	0.519	0.49	0.60	0.0113	0.0005	5.20	0.0076	0.0062	0.0021
A6	VD-end	0.539	0.46	0.60	0.0112	0.0004	5.20	0.0063	0.0012	0.0022
A7	LF3-end	0.538	0.47	0.60	0.0113	0.0004	5.25	0.0067	0.0017	0.0017
A8	VC- initial	0.536	0.47	0.60	0.0117	0.0006	5.27	0.0062	0.0006	0.002

Table 2. Main chemical composition of slag specimens in mass %.

Specimen No.	Stage	CaO	SiO_2	Al_2O_3	MnO	Cr_2O_3	FeO	CaF_2
A1	IF-end	5.91	46.30	19.90	13.30	6.08	3.51	0.17
A2	LF1-initial	6.62	42.20	19.10	11.70	7.19	3.63	0.32
A3	LF1-end	63.17	19.40	6.48	0.16	0.14	0.48	3.86
A4	LF2-initial	—	—	—	—	—	—	—
A5	LF2-end	61.69	19.30	4.55	0.03	0.04	0.25	8.07
A6	VD-end	58.91	20.80	6.51	0.02	0.03	0.16	7.06
A7	LF3-end	58.25	19.10	6.99	0.04	0.05	0.38	7.98
A8	VC- initial	57.83	19.10	6.50	2.30	0.05	0.35	7.59

3. Results

3.1. Analysis of Defective Part on the Ingot

The defective parts that failed the magnetic particle inspection test were investigated through SEM and EDS analysis method, which is shown in Figure 3. A linear and aggregating defect up to 3 mm was observed, which could lead to the failure of flaw detection directly, as shown in Figure 3a. Based on the element mapping of defects, the linear and aggregating defect is composed of single SiO_2-MnO-Al_2O_3 inclusion, which presented irregular in shape and uneven in size, as shown in Figure 3b. The morphology and element mapping of single and large-size inclusions are shown in Figure 3c,d, respectively. The large SiO_2-MnO-Al_2O_3 inclusion that demonstrated in irregular plate strip is up to nearly 372 μm. In general, the SiO_2-MnO-Al_2O_3 type inclusions considered as easily deformable have a low liquidus temperature and contact angle, which could increase difficulties of the removal by its floatation [31–33]. Besides, large-size inclusions, e.g., SiO_2-MnO-Al_2O_3, CaO-SiO_2-Al_2O_3 inclusions with low contact angles were also considered to have a poor tendency of agglomeration [34]. However, Vantilt et al. [32] found the cluster and agglomeration of liquid, spherical SiO_2-MnO-Al_2O_3 inclusions driven by the capillary depression forces. The morphological difference of SiO_2-MnO-Al_2O_3 inclusions from spherical to irregular was suggested as the result of forging. In summary, it could be suggested that the formation of the large-size and aggregating SiO_2-MnO-Al_2O_3 type inclusions led to the failure in the previous MPI test.

3.2. Characteristics of the Inclusions during the Metallurgical Process

To clarify the origin and formation of large-size SiO_2-MnO-Al_2O_3 type inclusions in molten steel, the evolution of inclusions during the metallurgical process was investigated through SEM and EDS analysis, especially for the large-size inclusions. Morphology and composition of typical large-size inclusions at different stages were shown in Table 3. Typical inclusions in specimen A1 (IF-end before tapping) were large-size spherical SiO_2-MnO-Al_2O_3 inclusions up to 30 μm, which were even-distributed in the composition based on the micrograph. After the tapping, the main inclusion type remained as SiO_2-MnO-Al_2O_3. In the meanwhile, a few CaO content appeared in some SiO_2-MnO-Al_2O_3 inclusions, and the shapes changed from sphere to non-smooth and irregular sphere. After the addition of Si-Al complex deoxidizer during the following LF1 and LF2 refining process, the main inclusions converted into CaO-SiO_2-Al_2O_3 type with a low liquidus temperature and contact angle as well [31,34]. Additionally, only a few SiO_2-MnO-Al_2O_3 inclusions with decreasing sizes were found at the end of the LF2 refining process. After the breaking of VD, the CaO- SiO_2- Al_2O_3 inclusions were observed as a minor part. The main inclusion changed from CaO-SiO_2-Al_2O_3 to SiO_2-MnO-Al_2O_3 type. In the following LF3 refining and VC casting process, the main inclusion remained still as CaO-SiO_2-Al_2O_3. In the meanwhile, SiO_2-MnO-Al_2O_3 type became minor with decreasing sizes gradually.

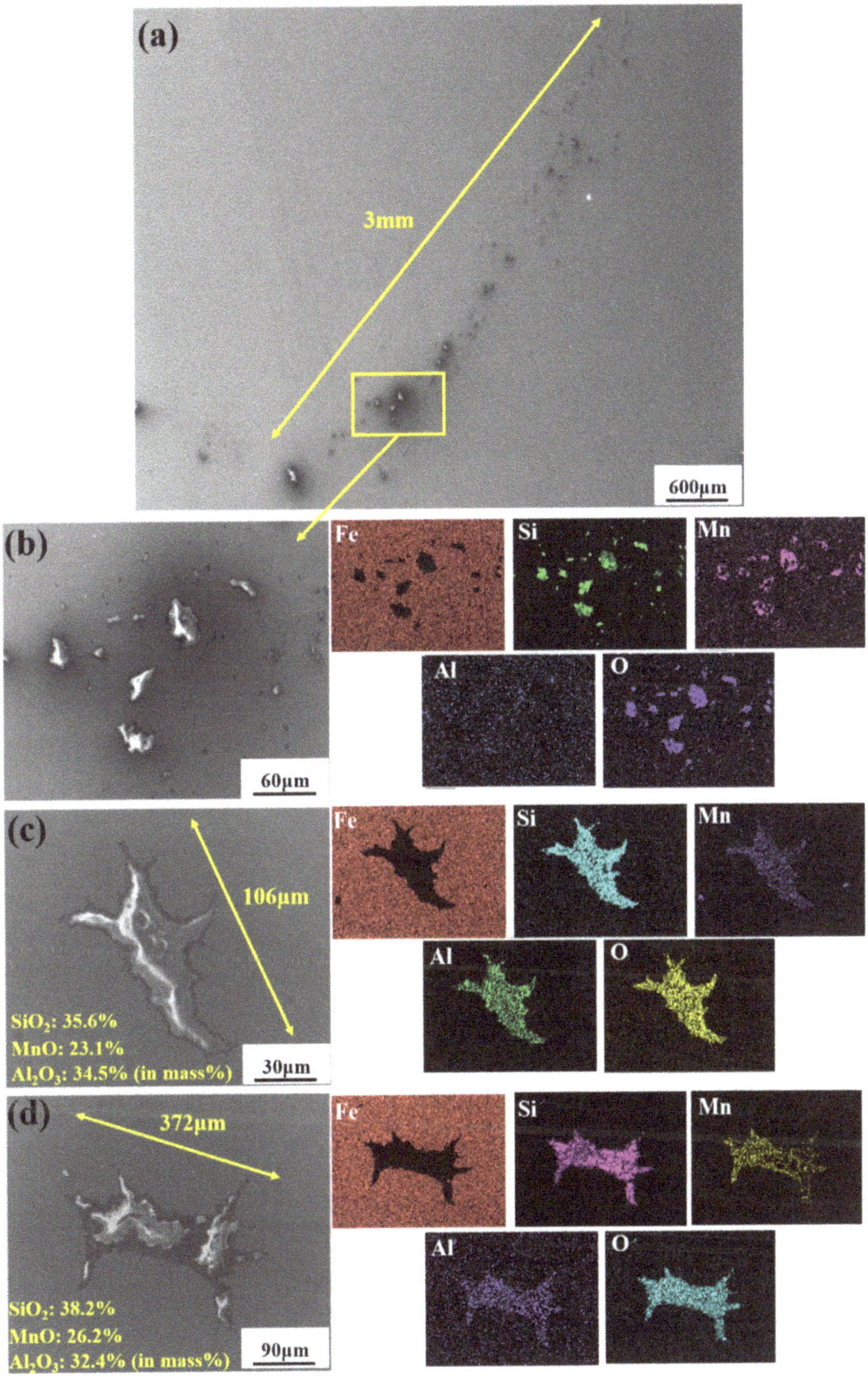

Figure 3. Morphology and element mapping of the defective area on the near surface of ingot. (**a**) morphology of defects in overall; (**b**) clusters of SiO_2-MnO-Al_2O_3 inclusions; (**c**,**d**) large-size single SiO_2-MnO-Al_2O_3 inclusion.

Table 3. Morphology and composition (in mass%) of typical large-size inclusions at different stages (Specimen A1–A8).

Sample No.	Stage	Main Inclusions		Minor Inclusions	
		Morphology	**Type wt.%**	**Morphology**	**Type wt.%**
A1	IF-end	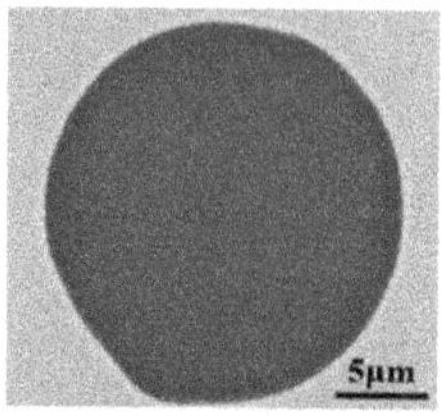	SiO_2-(48.0%) MnO-(22.3%) Al_2O_3-(29.6%)	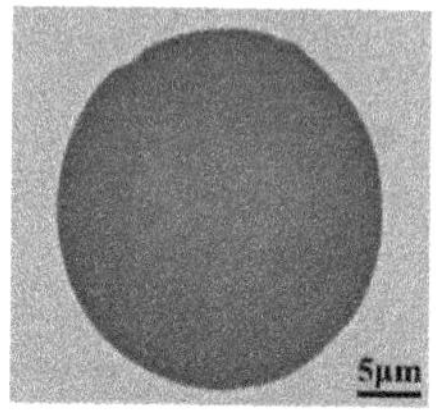	SiO_2-(72.4%) MnO-(18.6%) Al_2O_3-(0.09%)
A2	LF1-intial	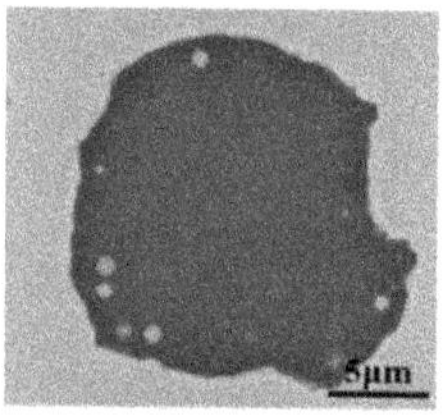	SiO_2-(28.2%) MnO-(14.2%) Al_2O_3-(46.5%) CaO-(10.5%)	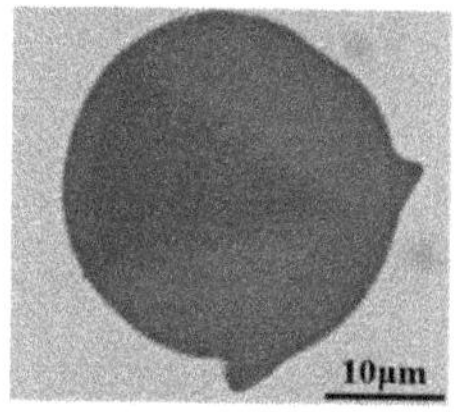	SiO_2-(58.1%) MnO-(12.3%) Al_2O_3-(29.6%)
A3	LF1-end		CaO-(11.2%) SiO_2-(49.3%) Al_2O_3-(39.5%)	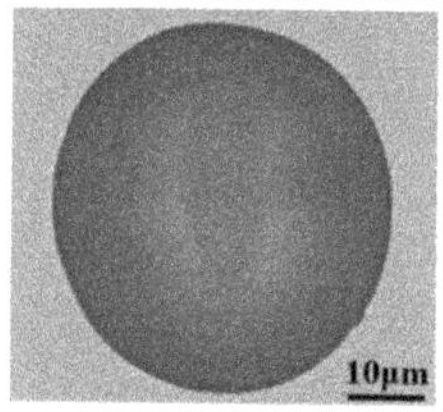	CaO-(50.7%) SiO_2-(33.6%) Al_2O_3-(15.7%)
A4	LF2- initial	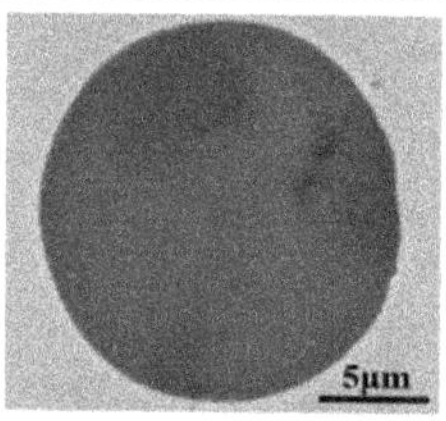	CaO-(24.4%) SiO_2-(24.4%) Al_2O_3-(51.2%)	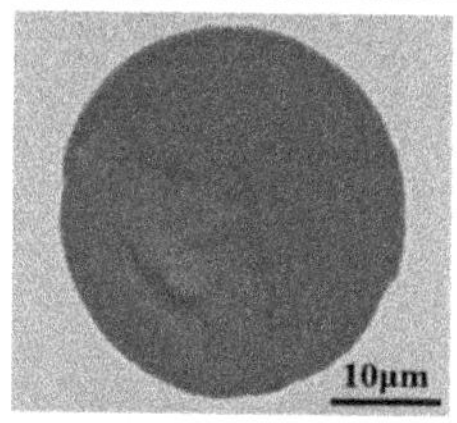	CaO-(36.2%) SiO_2-(26.9%) Al_2O_3-(36.9%)
A5	LF2-end	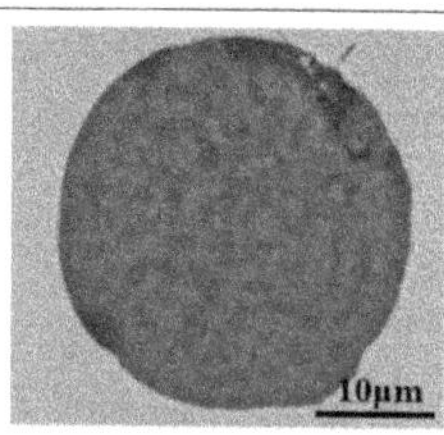	CaO-(79.0%) SiO_2-(10.0%) Al_2O_3-(11%)	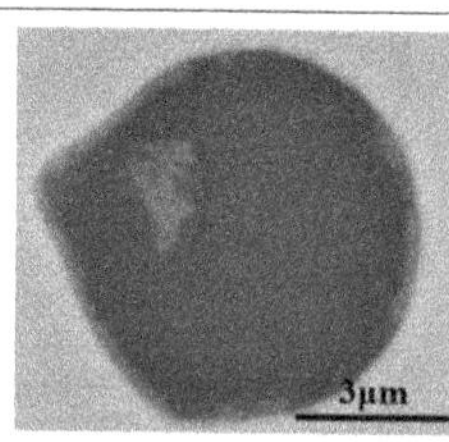	SiO_2-(42.0%) MnO-(22.5%) Al_2O_3-(35.0%)
A6	VD-end	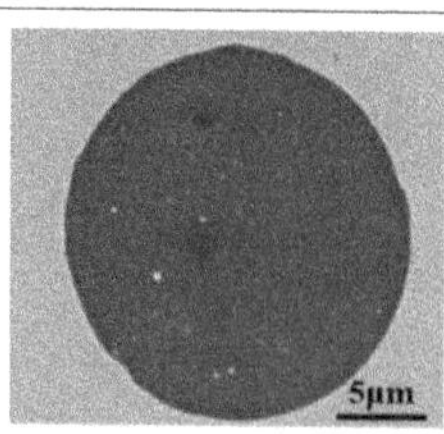	SiO_2-(72.5%) MnO-(22.3%) Al_2O_3-(5.1%)	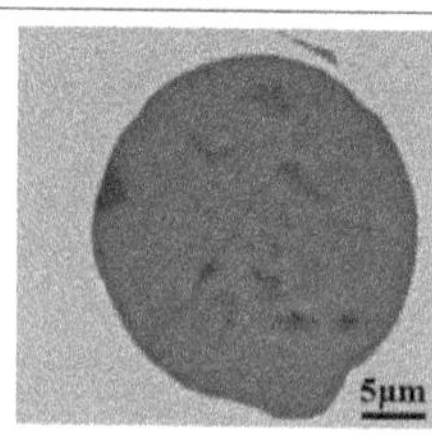	CaO-(75.7%) SiO_2-(18.5%) Al_2O_3-(5.9%)

Table 3. *Cont.*

Sample No.	Stage	Main Inclusions		Minor Inclusions	
		Morphology	Type wt.%	Morphology	Type wt.%
A7	LF3-end	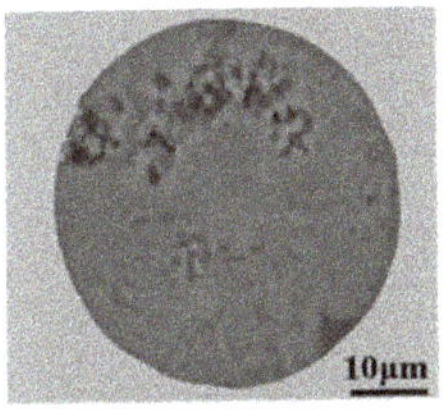	CaO-(63.8%) SiO_2-(18.2%) Al_2O_3-(18.0%)	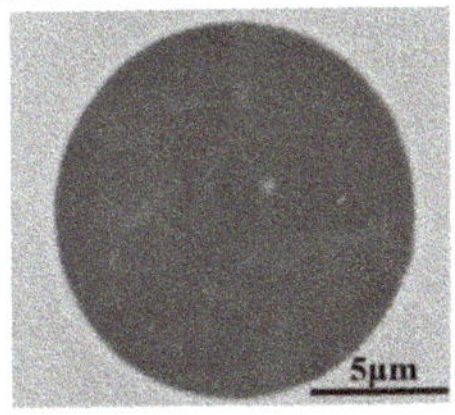	SiO_2-(25.0%) MnO-(22.3%) Al_2O_3-(54.2%)
A8	VC-initial	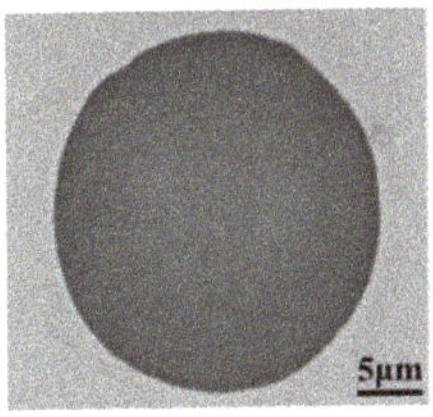	CaO-(52.6%) SiO_2-(12.6%) Al_2O_3-(34.8%)	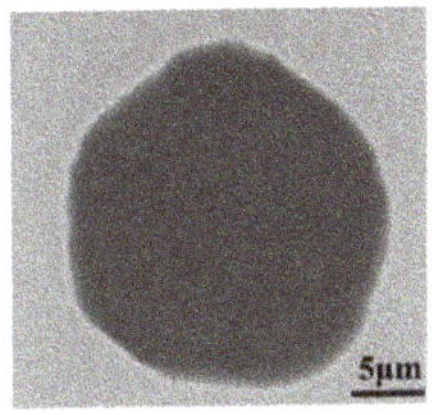	SiO_2-(15.0%) MnO-(12.3%) Al_2O_3-(72.5%)

To illustrate the evolution of large-size inclusions at a different stage, the composition and size of inclusions found in specimen A1–A8 were marked in the ternary diagram of CaO-SiO_2-Al_2O_3 and SiO_2-MnO-Al_2O_3 system utilizing Factsage 8.1 thermodynamics software, (version 8.1, CRCT, Montreal, Canada) as shown in Figure 4a–d, respectively. The isothermal temperature of each diagram was set as 1873 K. The red line represented the liquid oxide phase region. During the smelting process, a few inclusions are located in the liquid oxide phase. Inclusions were mainly distributed around CaO and SiO_2 corners in the CaO-SiO_2-Al_2O_3 ternary diagram shown in Figure 4a. At the end of the IF melting process, the composition of inclusions was mainly located in the SiO_2 corner with a certain amount of Al_2O_3. During the LF1 refining process, the inclusions were going to own higher Al_2O_3 and CaO contents gradually. Then, most inclusions found in the LF2-initial stage were mainly liquid oxide. However, the CaO content in the inclusions compared with the initial stage has risen rapidly till the end of LF2-end. Despite the slight decrease of CaO content in a few inclusions, however, the main inclusions with larger sizes were still found around the corner of SiO_2. Afterward, the composition of inclusions stayed roughly stable in the following process. Due to the limits of the ternary diagram, the composition of inclusions was also projected into the SiO_2-MnO-Al_2O_3 system to clarify the evolution of SiO_2-MnO-Al_2O_3 type inclusions, which were shown in Figure 4b,d. SiO_2-MnO-Al_2O_3 type inclusions were mainly found at the initial stage before the LF1 refining process. Meanwhile, the composition of SiO_2-MnO-Al_2O_3 inclusions found after the breaking of VD were similar to the ones at the IF-end. Moreover, the composition of SiO_2-MnO-Al_2O_3 in the defective area marked in Figure 4b was located in the liquid oxide phase.

The size of inclusions was represented by the diameter of circles in the diagram as shown in Figure 4c,d, respectively, which were mainly spherical both in the CaO-SiO_2-Al_2O_3 and SiO_2-MnO-Al_2O_3 system. During the latter half of LF1 and the whole LF2 refining process, the sizes of inclusions were up to 30 μm owing to the collision, agglomeration, floatation, and absorption by high basicity slag. Super large-size inclusion up to 100 μm was only observed at the breaking of VD. The rest of the large-size inclusions were mainly located in dual-phase regions, which were around the corner of CaO and SiO_2. However, the sizes of inclusions found in the LF3-end and VC-initial were mostly rather smaller compared with VD-end ones. In addition, the sizes of pure liquid inclusions were relatively smaller. The SiO_2-MnO-Al_2O_3 inclusions were firstly observed at the stage of IF-end, then grew up bigger during the stage of the LF refining process. A huge amount of large-sized SiO_2-MnO-Al_2O_3 inclusions that were unabsorbed and undissolved in the refining slag were found after the breaking of VD, which could lead to the failure of flaw detection directly.

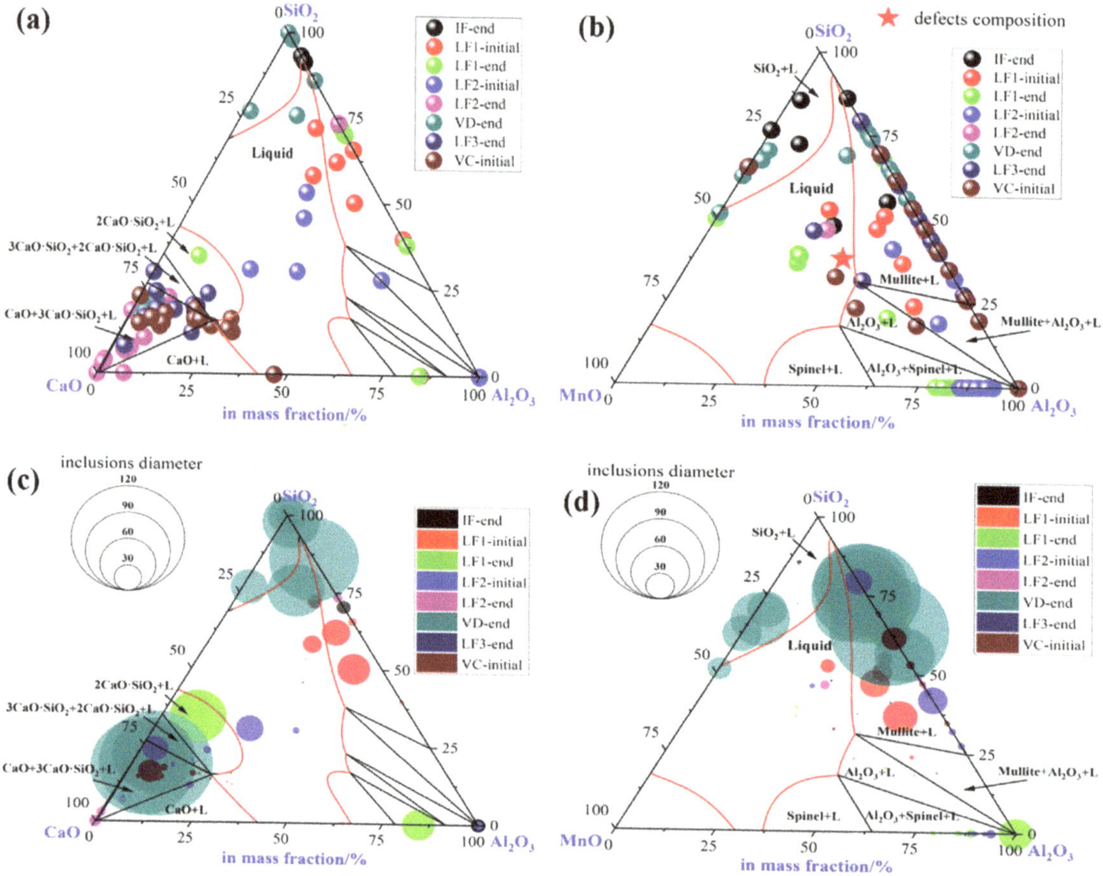

Figure 4. Composition and size distribution of inclusions during the metallurgical process in CaO-SiO_2-Al_2O_3 and SiO_2-MnO-Al_2O_3 ternary phase diagram at 1873 K. (**a**) Composition of inclusions in CaO-SiO_2-Al_2O_3 diagram, (**b**) composition of inclusions in SiO_2-MnO-Al_2O_3 diagram, (**c**) size distribution of inclusions in CaO-SiO_2-Al_2O_3 diagram, (**d**) size distribution of inclusions in SiO_2-MnO-Al_2O_3 diagram.

The type and the average size of all type inclusions at each stage was shown in Figure 5. The SiO_2-MnO-Al_2O_3 inclusions were found during the whole metallurgical process. The occurrence of Al_2O_3 and CaO-SiO_2-Al_2O_3 inclusions were presented as complementary. Owing to the incomplete deoxidation in the molten steel and stirring between slag and molten steel during the avoiding of the top slag process, the Al_2O_3 inclusions were found mostly before the LF2 refining process. Then the Al_2O_3 inclusions were modified into the MgO-Al_2O_3 spinel inclusions due to the dissolution of MgO-based refractory. The CaO-SiO_2-Al_2O_3 inclusions remained as the majority till the VC-initial stage. The average size of inclusions at the breaking of VD was much higher than the other stages owing to the appearance of the large amount, large-sized SiO_2-MnO-Al_2O_3 inclusions.

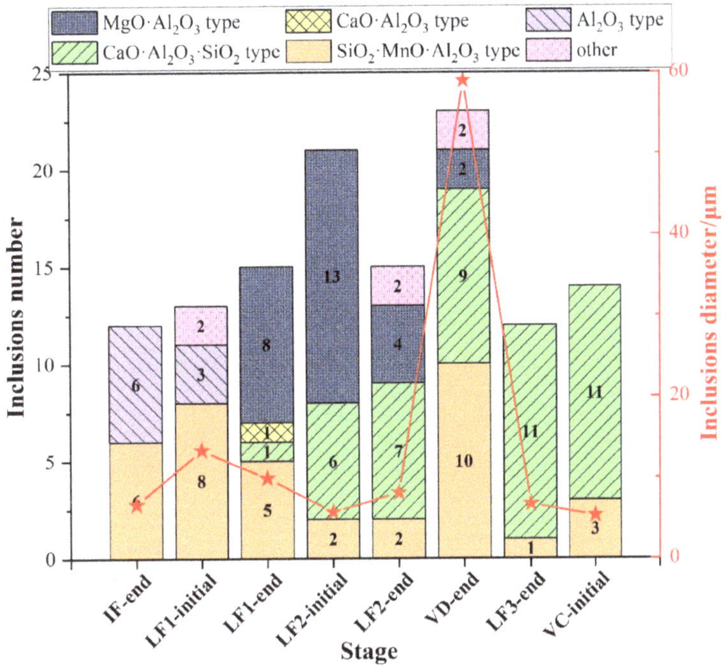

Figure 5. Type and average size distribution of all inclusions at different stages.

4. Discussion

4.1. Formation Mechanism of the SiO_2-MnO-Al_2O_3 Type Inclusions

Based on observation and analysis for the inclusions found in the metallurgical process, it could be inferred that the SiO_2-MnO-Al_2O_3 inclusions were formed at the end of IF, and remained in the following stages. To clarify the formation mechanism of the SiO_2-MnO-Al_2O_3 inclusion, the stability phase diagrams of the Si-Mn-O system in the Fe-5Cr back-up roller system were calculated through Factsage 8.1 software at different aluminum content (0, 0.003%, 0.007%, 0.01%, 0.015%). The colored dotted line represented different oxygen content in the molten steel, which was shown in Figure 6a–e, respectively. As shown in Figure 6a, at the 0 aluminum content, it could be found that liquid oxide, SiO_2, spinel (MnO-Cr_2O_3), and MnO phases in the molten steel with the content of silicon and manganese ranging from 1 ppm to 10% and 100 ppm to 10%, respectively. Besides, the regions of SiO_2, MnO, and liquid oxide phase became larger with the increasing oxygen content in the molten steel, which also indicated that the increasing degree of over-oxidation extended the formation scale of liquid oxide. The spinel and SiO_2 phase-field were modified into Al_2O_3 and mullite phase fields. The Al_2O_3+liquid oxide dual-phase field was formed in the previous pure liquid oxide phase-field region when the aluminum content increased to 30 ppm. Due to the similar aluminum content, the composition of A3 and A4, A5, and A6 were located in the dual-phase of Al_2O_3+liquid oxide and marked in Figure 6b,c, respectively. It could be indicated that the endogenous inclusions of each stage should be Al_2O_3+liquid oxide type, which was verified above. Due to the strong stirring of the VD process and the similar SiO_2-MnO-Al_2O_3 inclusions composition between IF-end and VD-end, as shown in Figure 5b, the SiO_2-MnO-Al_2O_3 inclusions found in the A6(VD-end) could be suggested as the result of slag entrapment containing unabsorbed or undissolved, large-sized SiO_2-MnO-Al_2O_3 inclusions. At a certain oxygen content in the molten steel, the region related to the liquid oxide became smaller with the increasing aluminum content. Consequently, it could be inferred that the aluminum content could promote the decrease of liquid oxide effectively.

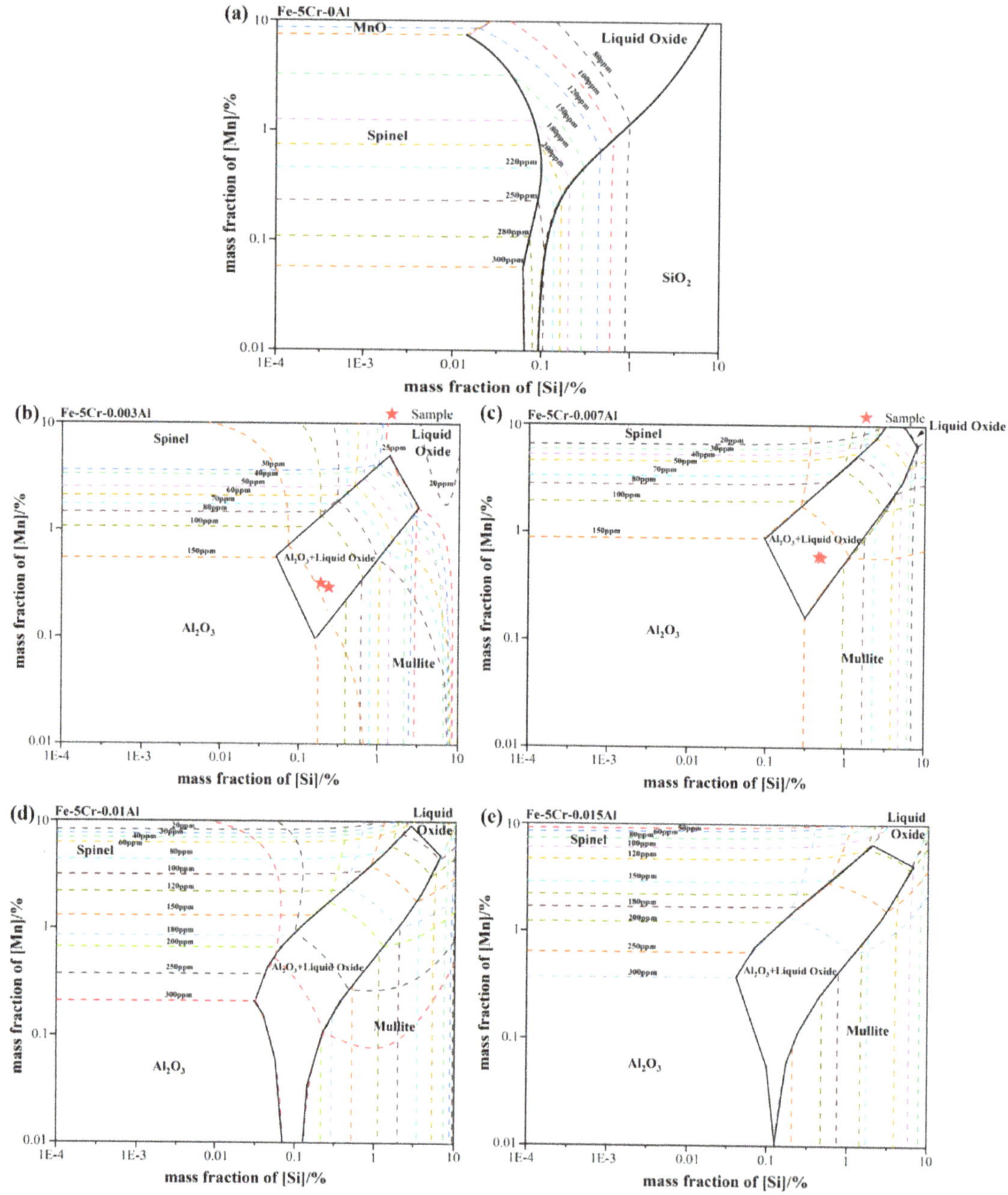

Figure 6. Calculated oxide stability diagrams of Si-Mn-O system with iso-oxygen contours (in ppm) in Fe-5Cr back-up roller at 1873 K: (**a**) with 0 Al (**b**) with 30 ppm Al (**c**) with 70 ppm Al (**d**) with 0.01% Al (**e**) with 0.015% Al.

4.2. Modification of the Liquid Oxide Inclusions

To obtain the quantitative relationship between the aluminum content and the formation of liquid oxide, the evolution of oxide inclusions in Fe-5Cr-0.55C-0.55Si-0.6Mn-6 ppm Ca containing different aluminum contents was calculated with the aid of Factsage

8.1 software, as shown in Figure 7. The FactIPS, FTmisc, and FToxide databases were used for calculation. The soluble oxygen contents in the molten steel ranged from 20 ppm to 200 ppm were shown in Figure 7a,c,e,g, respectively, which corresponded to the different stages in the smelting process. The upper and lower aluminum contents during the smelting process were marked in a blue dotted line. As shown in Figure 7g, at the oxygen content of 200 ppm in molten steel, the SiO_2, liquid oxide, and Al_2O_3 phase existed when the aluminum content was below 100 ppm corresponded to the actual value. The liquid oxide phase disappeared completely, which was replaced by the CA_6 (CaO-$6Al_2O_3$) phase till the aluminum content reached roughly 0.024%. Besides, the SiO_2, Al_2O_3, and MnO contents in the liquid oxide stayed as the dominating part, and the CaO content in liquid oxide stayed as a minor part. Then, the liquid oxide formed at 200 ppm oxygen content was determined as SiO_2-MnO-Al_2O_3 type inclusions which corresponded to the observed results above. Consequently, it could be inferred that a minimum of 0.024% aluminum content in the molten steel was required to convert the SiO_2-MnO-Al_2O_3 type liquid oxide into the high melting point Al_2O_3 inclusions, which were widely recognized as easier to remove by their floatation and slag absorption than liquid oxides did [34–37]. Generally, the contact angles of liquid oxide and solid oxide are below 90° and beyond 90° at 1873 K, respectively, which could reflect the difficulties of slag absorption. The liquid film was formed between the liquid inclusions and slag interface when the liquid inclusions were about to go through the slag-steel interface. Then the slow discharge of liquid film would take rather longer to get absorbed than the solid ones, in which no liquid film has been formed during the process. Then, it could be concluded that reducing of liquid oxide (SiO_2-MnO-Al_2O_3 and CaO-SiO_2-Al_2O_3 type) and modification to the Al_2O_3 type solid inclusions could promote the cleanness of molten steel effectively. The aluminum content required for the complete modification of solid Al_2O_3 type inclusions reduced from 0.024% to 0.01% when the oxygen level declined from 200 ppm to 50 ppm, which corresponded to the process of IF melting to LF refining, as shown in Figure 7c,d,e,f. However, in the meanwhile, the CaO content in liquid oxide has risen higher than MnO at the 50 ppm oxygen content in molten steel, which meant the newly formed liquid oxide transformed from SiO_2-MnO-Al_2O_3 type to CaO-SiO_2-Al_2O_3 type corresponding to the analysis in the LF2-refining process. In addition, the newly formed stable phases were converted to CA_2 (CaO-$2Al_2O_3$) and CA6 (CaO-$6Al_2O_3$) phases by the effect of high basicity slag. The required aluminum content for the complete modification rose to 0.02% at the 20 ppm oxygen content in molten steel, which represented the end of the LF refining process in the plant trial. Due to the modification of the high basicity slag and complete oxidation in molten steel, the solid inclusions were converted into pure CA_2 inclusion. The content of CaO in liquid oxide has risen higher than SiO_2 gradually, as shown in Figure 7a,b. Above all, it could be suggested that a minimum of 0.024% aluminum content could minimize the formation of SiO_2-MnO-Al_2O_3 liquid oxide before the tapping. Besides, maintaining the aluminum content beyond 0.02% in the whole refining process could also minimize the formation of CaO-SiO_2-Al_2O_3 liquid oxide.

4.3. Optimization and Verification of Plant Trials

Based on the results and analysis above, the authors brought the aluminum control over the whole smelting process into effect to solve the flaw detection issues caused by SiO_2-MnO-Al_2O_3 type inclusions, as shown in Figure 8. Meanwhile, the composition of slag was redesigned to match both the high aluminum content in molten steel and the refining function. Then, the steel specimens corresponding to the previous stages were also sampled and analyzed as above. Specific details of sampling in the optimized plant trial are shown in Figure 8. The composition of slag specimens in the optimized process is shown in Table 4.

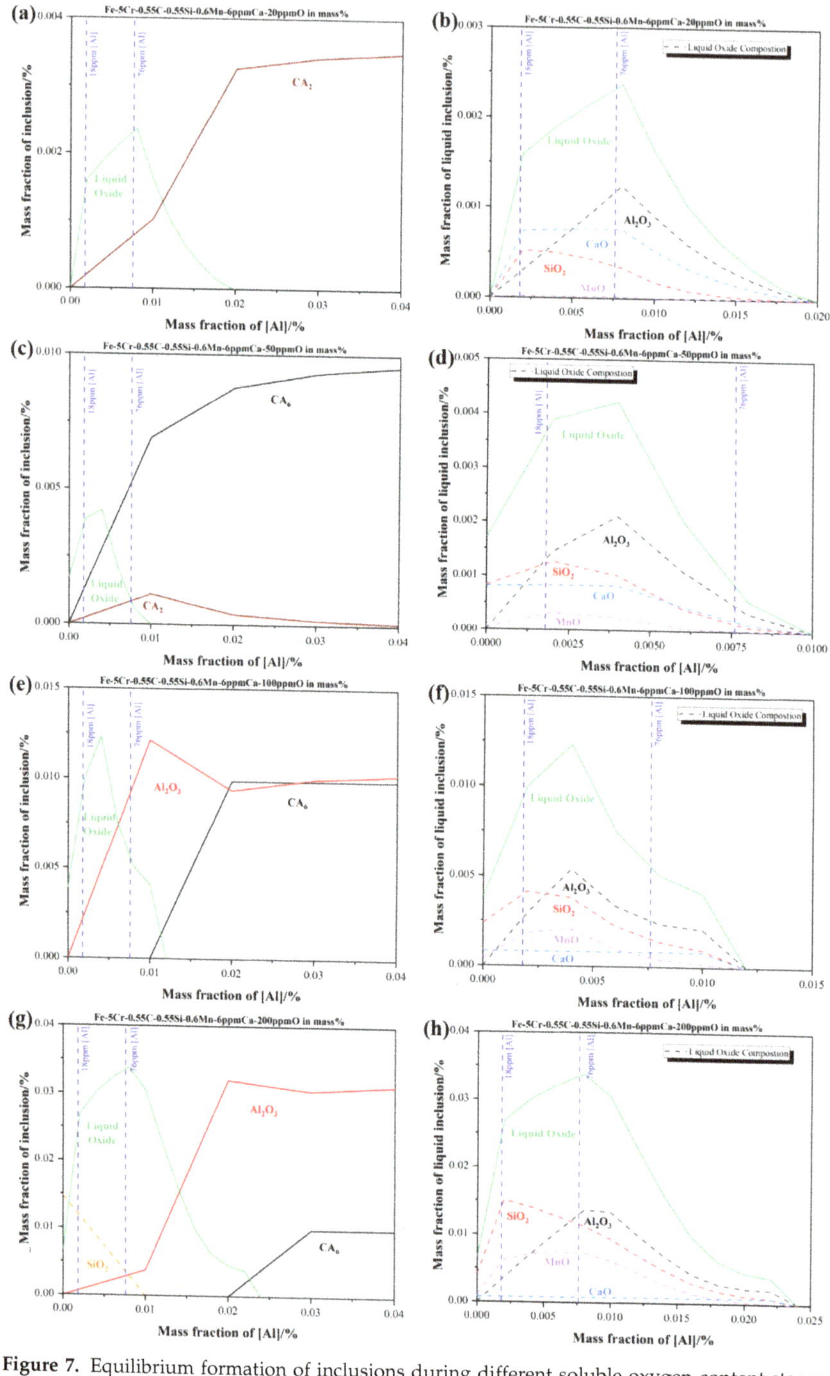

Figure 7. Equilibrium formation of inclusions during different soluble oxygen content stages at 1873 K in Fe-5Cr-0.55C-0.55Si-0.6Mn-6 ppm Ca-Al in mass % (**a**,**b**) with 20 ppm (**c**,**d**) with 50 ppm (**e**,**f**) with 100 ppm (**g**,**h**) with 200 ppm (**a**,**c**,**e**,**g**) present the formation of all inclusions at different oxygen content (**b**,**d**,**f**,**h**) present the actual composition of liquid oxide inclusions corresponding to the previous one, respectively.

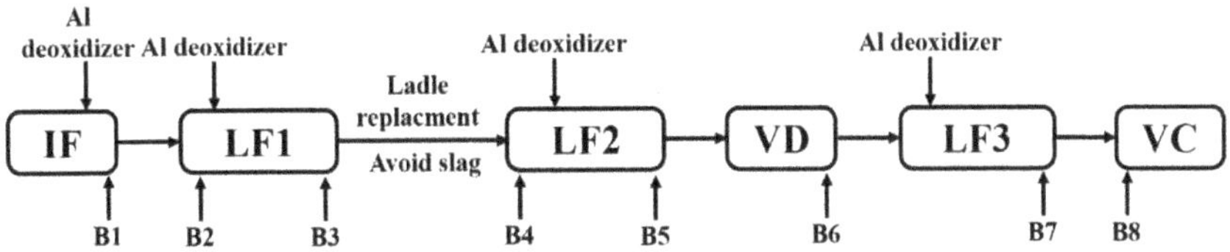

Figure 8. Schematic illustration of optimized metallurgical process and sampling.

Table 4. Main chemical composition of slag specimens in the optimization process in mass%.

Specimen No.	Stage	CaO	SiO_2	Al_2O_3	MnO	Cr_2O_3	FeO	CaF_2
B1	IF-end	0.99	41.00	19.00	20.40	10.00	3.82	0.00
B2	LF1-initial	2.36	31.80	36.50	13.60	6.51	3.90	0.08
B3	LF1-end	62.79	16.70	10.60	0.20	0.15	0.57	4.19
B4	LF2-initial	—	—	—	—	—	—	—
B5	LF2-end	54.39	12.00	23.20	0.15	0.20	1.07	2.24
B6	VD-end	53.07	11.30	24.50	0.05	0.10	0.47	2.13
B7	LF3-end	53.33	11.10	24.80	0.05	0.10	0.46	1.97
B8	VC- initial	52.88	11.00	25.20	0.07	0.12	0.80	2.01

Morphology and composition of improved typical inclusions at different stages were shown in Table 5. Only aggregating and irregular Al_2O_3 particles were found before the LF1 refining process. Then, pure Al_2O_3 particles were modified into small-sized and quasi-spherical CaO-Al_2O_3 type inclusions in the LF refining process. However, the CaO-Al_2O_3 type inclusions were modified into CaO-SiO_2-Al_2O_3 type inclusions after the breaking of the VD process. The CaO-SiO_2-Al_2O_3 type inclusions remained as the main inclusions till the casting of the VC process. To illustrate the optimized evolution of large-size inclusions at different stages, the composition and size of inclusions found in specimens B1-B8 were marked in the ternary diagram of the CaO-SiO_2-Al_2O_3 system, as shown in Figure 9. During the whole smelting process, a few inclusions were located at the line of SiO_2-Al_2O_3, which meant few SiO_2-MnO-Al_2O_3 type inclusions were found. Large-sized inclusions were mostly located in the liquid oxide and liquid oxide+CaO dual-phase field, which was relatively smaller than the SiO_2-MnO-Al_2O_3 type ones. Besides, large-sized inclusions were CaO-Al_2O_3 type at the end of LF2 refining. However, the liquid oxides were mostly converted to CaO-SiO_2-Al_2O_3, which were formed after the breaking of VD. It could be suggested that the inclusions with the tendency of high Al_2O_3 content usually owned small sizes. On the contrary, the inclusions with the tendency of high CaO content or a certain SiO_2 content owned larger sizes generally in this study. Fewer inclusions were found than the original process especially the SiO_2-MnO-Al_2O_3 type, which caused the failure of the MPI test. It showed that the modification of aluminum in the over-oxidation environment before the tapping was effective in a certain degree. In addition, the amount and sizes of inclusions in the optimized process were smaller than the previous ones.

Table 5. Morphology and composition (in mass %) of typical inclusions at different stages (Specimen B1–B8).

Specimen No.	Stage	Main Inclusions	
		Morphology	Type wt.%
B1	IF-end	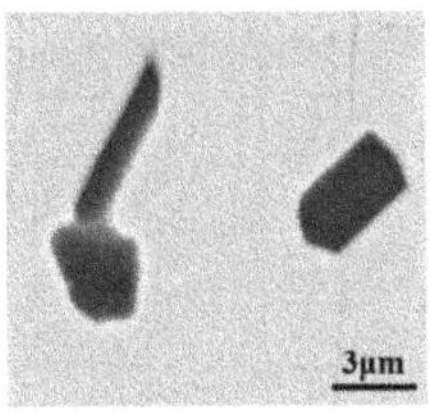	Al_2O_3

Table 5. *Cont.*

Specimen No.	Stage	Main Inclusions	
		Morphology	Type wt.%
B2	LF1-initial	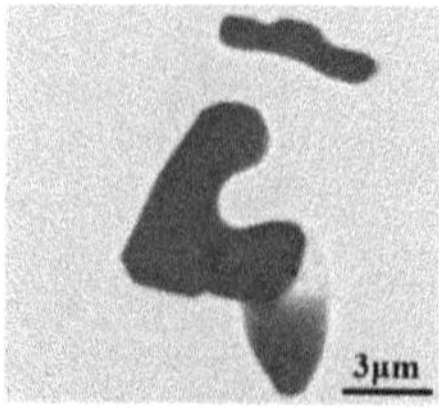	Al_2O_3
B3	LF1-end	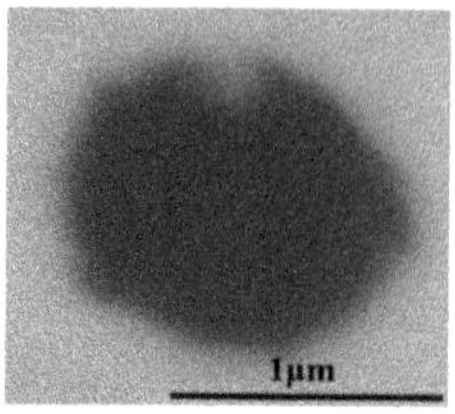	CaO-(11.2%) Al_2O_3-(83.2%)
B4	LF2-initial	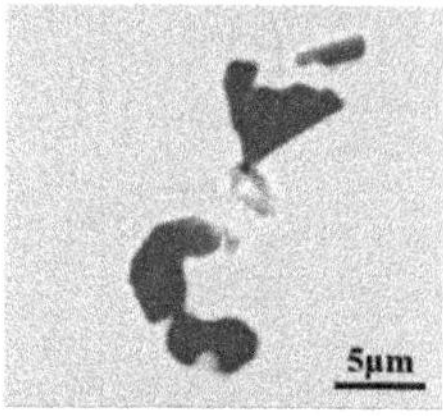	Al_2O_3
B5	LF2-end	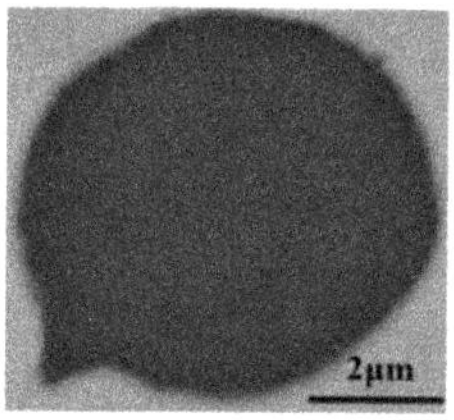	CaO-(57.6%) Al_2O_3-(40.2%)
B6	VD-end	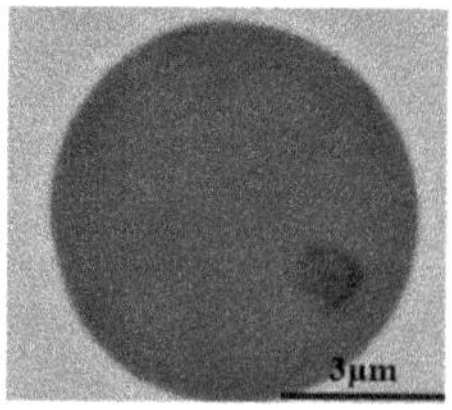	CaO-(66.5%) SiO_2-(8.6%) Al_2O_3-(25.9%)
B7	LF3-end	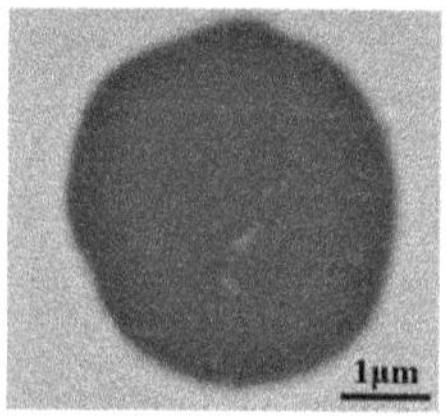	CaO-(55.8%) SiO_2-(9.2%) Al_2O_3-(36.0%)

Table 5. *Cont.*

Specimen No.	Stage	Main Inclusions	
		Morphology	Type wt.%
B8	VC-initial	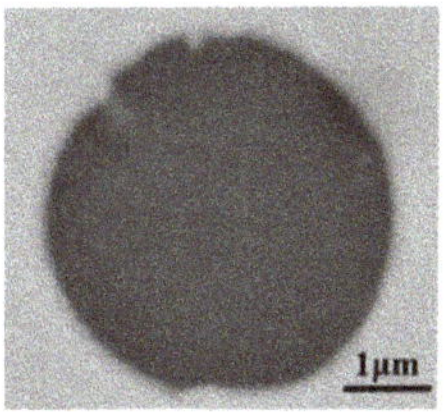	CaO-(40.9%) SiO_2-(6.4%) Al_2O_3-(52.7%)

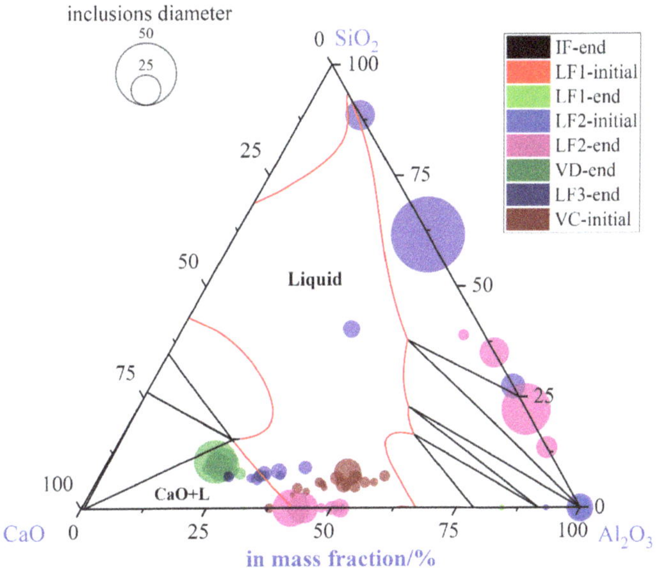

Figure 9. Composition and size distribution of large-size inclusions during the optimized metallurgical process in CaO-SiO_2-Al_2O_3 ternary phase diagram at 1873 K.

The type and the average size of all type inclusions at each stage was shown in Figure 10. SiO_2-MnO-Al_2O_3 inclusions were found after the LF2-initial stage, which could be entrapped from the slag during the avoiding slag process. Owing to the optimized control of aluminum content, Al_2O_3 inclusions dominated in inclusions before LF-2 refining, which were modified by the calcium content in the high basicity slag. Besides, the inclusion found at the breaking of VD owned larger sizes than other stages. The appearance of Al_2O_3 and CaO-SiO_2-Al_2O_3 inclusions was also presented as complementary, which matched the previous results. However, more CaO-SiO_2-Al_2O_3 inclusions were found after the breaking of the VD process than the previous ones. Therefore, the relationship between SiO_2, Al_2O_3 content in the inclusions, and the aluminum content in the molten steel was illustrated and shown in Figure 11. The SiO_2 contents in the inclusions and the aluminum content in the molten steel presented a negative correlation. At the end of LF2 refining, the SiO_2 contents in the inclusions were almost none. However, the Al_2O_3 contents of the inclusions decreased rapidly when the smaller SiO_2 contents increased. The SiO_2 contents decreased gradually after adding in the aluminum wire (aluminum 99%). Consequently, it could be inferred that the aluminum contents in the molten steel could suppress the formation of SiO_2 contents in the CaO-SiO_2-Al_2O_3 inclusions to a certain degree. However, the precise inner mechanism requires further investigation [38].

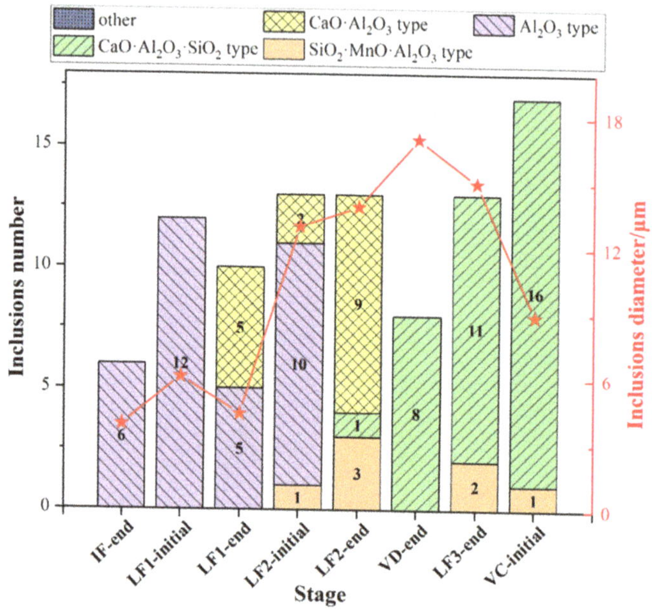

Figure 10. Type and average size distribution of all inclusions at different stages in the optimized plant trial.

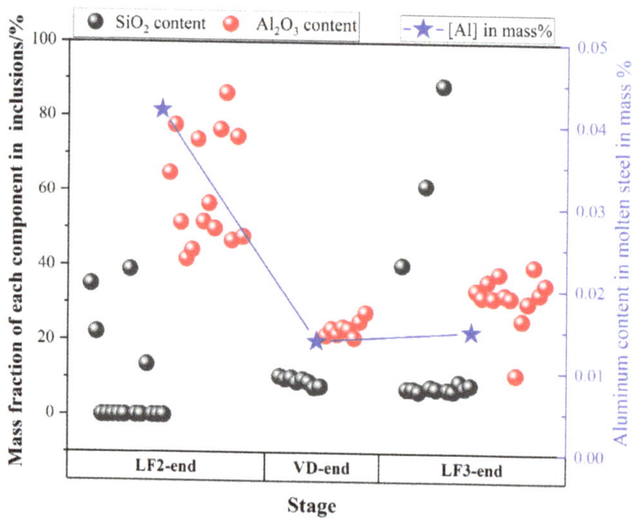

Figure 11. Mass fraction of SiO_2 and Al_2O_3 contents in the inclusions and the corresponding aluminum content during LF2-end, VD-end, LF3-end stage.

According to the experimental results and thermodynamic calculation, the formation and evolution mechanism of typical inclusions during the original and optimized metallurgical process were illustrated and shown in Figure 12. The scarps and alloys were melted by IF under the atmosphere. Then there would be a huge amount of FeO, MnO, and Cr_2O_3 oxide formed and entered into the top slag, which provided the oxygen content into the molten steel. If there were insufficient aluminum content in the molten steel, the huge number of SiO_2-MnO-Al_2O_3 inclusions would be formed by the joint control of silicon, manganese, and aluminum over the oxygen content in the molten steel, which was not easy to remove by its floatation and the absorption of the slag. In the refining stage, the CaO content would emerge in both the SiO_2-MnO-Al_2O_3 and Al_2O_3 inclusions by the modification of the high basicity slag. Besides, the CaO-Al_2O_3 type inclusions were converted into CaO-SiO_2-Al_2O_3 inclusions owing to the insufficient aluminum content caused the aluminum loss during the VD process and the slag entrapment [39–41]. The precise and inner mechanism of these requires further investigation in the next stage.

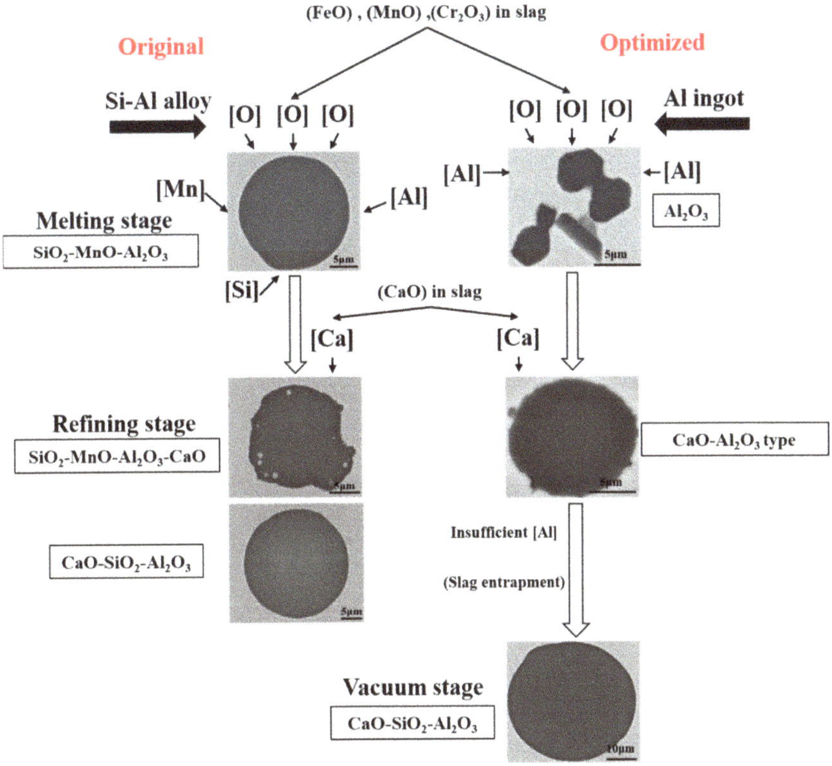

Figure 12. Schematic of formation and evolution of typical inclusions during the original and optimized metallurgical process.

5. Conclusions

In the current study, the source and formation of the defects that failed in the MPI test in forged backup rollers were investigated and the corresponding optimized process in the plant trials was conducted. Based on the analysis of inclusions characteristics and thermodynamic calculations, the conclusions were obtained as follows:

(1) The linear and aggregating distribution of SiO_2-MnO-Al_2O_3 inclusions up to 3 mm was the cause of the failure in flaw detection. The sizes of SiO_2-MnO-Al_2O_3 were rather uneven. Besides, SiO_2-MnO-Al_2O_3 inclusions with irregular plate strip shapes were uniformly distributed in composition.

(2) SiO_2-MnO-Al_2O_3 type inclusions were mainly formed in the over-oxidation environment during the IF melting. Owing to its low contact angles, large-size SiO_2-MnO-Al_2O_3 inclusions in the defective area were inherited from the initial smelting process. CaO-SiO_2-Al_2O_3 inclusions formed during the LF refining process were the dominating inclusions in the whole process.

(3) The content of aluminum in the molten steel over 0.024% in Fe-5Cr back-up roller steel could suppress the formation of SiO_2-MnO-Al_2O_3 liquid oxide during an over-oxidation environment. CaO-SiO_2-Al_2O_3 liquid oxide inclusions could be completely

modified as solid CaO-Al_2O_3 type inclusions over 0.02% aluminum content during the refining process. After the optimization, the amount and sizes of large-size liquid inclusions (>50 μm) decreased significantly before the VD refining.

Author Contributions: Conceptualization and writing, W.Z.; methodology, W.Z. and Y.Z.; software, W.Z. and G.W.; validation, W.Z., G.W. and Z.Z.; formal analysis, W.Z. and Y.Z.; investigation, W.Z.; supervision, Y.Z. and G.C.; project administration, Y.Z. and G.C. All authors have read and agreed to the published version of the manuscript.

Funding: This research was funded by the National Natural Science Foundation of China, grant numbers U196021 and 51874034.

Institutional Review Board Statement: Not applicable.

Informed Consent Statement: Not applicable.

Data Availability Statement: Not applicable.

Acknowledgments: The authors would like to express appreciation to the National Natural Science Foundation of China for the financial support, the State Key Laboratory of Advanced Metallurgy at the University of Science and Technology Beijing, and to Luoyang CITIC HIC Casting and Forging Co., Ltd., for academic and experimental assistance.

Conflicts of Interest: The authors declare no conflict of interest.

References

1. Yu, S.D.; Chen, H. Current Status and Several Problems of Roller Industry in China. *Iron Steel* **2007**, *42*, 1–6.
2. Montmitonnet, P.; Buessler, P. A Review on Theoretical Analyses of Rolling in Europe. *ISIJ Int.* **1991**, *31*, 525–538. [CrossRef]
3. Gonzalez, W.B.; Llano, J.; García, J. *Metallurgical Application to Work and Back Up Rolls for Hot & Cold Rolling of Flat Products*; Rolls for Flat Rolling of Steel: Lugones, Spain, 2007.
4. Mammeri, A.; Belzune, F.J.; Rodirguez, C.; Torre, M.; Poveda, S.; Garcia, J. Mechanical Properties of Chromium Steels for Back up Rolling Rolls. *Rev. Metal.* **2003**, *39*, 107–113. [CrossRef]
5. Nakagawa, Y.; Hashimoto, T.; Katayama, H.; Morikawa, H. Development of High-Cr Roll for Hot Strip Mill with Superior Resistance to Surface Deterioration and Spalling of Outer Shell. *Tetsu-to-Hagane* **1988**, *74*, 1993–2000. [CrossRef]
6. Shimizu, M.; Shitamura, O.; Matsuo, S.; Kamata, T.; Kondo, Y. Development of High Performance New Composite Roll. *ISIJ Int.* **1992**, *32*, 1244–1249. [CrossRef]
7. Di Schino, A.; Di Nunzio, P.E. Metallurgical Aspects Related to Contact Fatigue Phenomena in Steels for Back-Up Rolls. *Acta Metall. Slovaca* **2017**, *23*, 62–71. [CrossRef]
8. Jia, R.; Li, X.D.; Li, J.C.; Gu, K. Present Research Situation on Cr5 Series Large Back-up Roll Process in China. *Hot Work. Technol.* **2013**, *42*, 62–65.
9. Mccann, J. Overview of Work Rolls for Cold Rolling. *Ironmak. Steelmak.* **2000**, *27*, 15–18.
10. Kang, X.; Li, D.; Xia, L.; Campbell, J.; Li, Y. Development of Cast Steel Back-up Roll. *Int. J. Cast Met. Res.* **2006**, *19*, 66–71. [CrossRef]
11. Ichino, K.; Ishikawa, S.; Kataoka, Y.; Toyooka, T. Improvement of Hot Wear Characteristic of High Speed Tool Steel Roll by Increase in Cr and Mo Contents. *Tetsu-to-Hagane* **2003**, *89*, 680–685. [CrossRef]
12. Yu, H.; Ji, C.; Chen, B.; Wang, C.; Zhang, Y. Characteristics and Evolution of Inclusion Induced Surface Defects of Cold Rolled IF Sheet. *J. Iron Steel Res. Int.* **2015**, *22*, 17–23. [CrossRef]
13. Sano, Y.; Hattori, T.; Haga, M. Characteristics of High-Carbon High Speed Steel Rolls for Hot Strip Mill. *ISIJ Int.* **1992**, *32*, 1194–1201. [CrossRef]
14. Kobayashi, K. Technical View on Forged Steel Rolls in Japan. *Tetsu-to-Hagane* **1971**, *57*, 725–737. [CrossRef]
15. Chen, J.L. Development of Cr5 Forged Blank of Cold-Rollers. *Forg. Stamp. Technol.* **2006**, *31*, 10–12.
16. Zhou, Q.; Zhang, J.; Yin, Y.; Zhai, M. Characterization of Inclusions in Axle Steel by Ingot Casting. *Metall. Res. Technol.* **2019**, *116*, 501–508. [CrossRef]
17. Huang, H.G.; Du, F.S.; Zhang, F. FEM Analyses on Forming Mechanics of Inclusion Defects Inner Heavy Backup-Roll Forging. *China Mech. Eng.* **2009**, *20*, 477–481.
18. Dong, B.L. Cause Analysis on Exposed Inclusion and Steel Melting Process Optimization of Backup Roll. *Heavy Cast. Forg.* **2019**, *40*, 46–48.
19. Gotoh, K.; Okada, H.; Sasaki, T.; Koide, T. Effects of Roll Surface Deteriorations on Scale Defect in Hot Rolling. *Tetsu-to-Hagane* **1998**, *84*, 861–867. [CrossRef]
20. Takechi, H.; Namba, K.; Kawasaki, K.; Fujiwara, K. Evaluation of the Fatigue Damage of Rolls for Strip Mills below the Surface by X-ray Diffraction Method. *Tetsu-to-Hagane* **1979**, *65*, 2067–2075. [CrossRef]

21. Ray, A.K.; Mishra, K.K.; Chaudhary, P.N. Failure Analysis of Rolls of Cold Rolling Mill in Steel Plant. In *Failure Analysis, Proceedings of the Clinic on Failure Analysis, Jamshedpur, India, 18–19 February 1997*; NML: Jamshedpur, India, 1997; pp. 37–46.
22. Wang, R.; Bao, Y.; Li, Y.; Yan, Z.; Li, D.; Kang, Y. Influence of Metallurgical Processing Parameters on Defects in Cold-Rolled Steel Sheet Caused by Inclusions. *Int. J. Miner. Metall. Mater.* **2019**, *26*, 440–446. [CrossRef]
23. Wang, Q.; Li, Z.; Shi, Y.; Wang, L.; Liu, F. Interior Crack and Its Formation Mechanism in Overlaying Weld of Back-up Rolls. *Eng. Fail. Anal.* **2013**, *34*, 268–277. [CrossRef]
24. Jamil, M.; Khan, A.M.; Hegab, H.; Sarfraz, S.; Sharma, N.; Mia, M.; Gupta, M.K.; Zhao, G.; Moustabchir, H.; Pruncu, C.I. Internal Cracks and Non-Metallic Inclusions as Root Causes of Casting Failure in Sugar Mill Roller Shafts. *Materials* **2019**, *12*, 2474. [CrossRef] [PubMed]
25. Lu, J.; Cheng, G.; Wu, M.; Yang, G.; Che, J. Detection and Analysis of Magnetic Particle Testing Defects on Heavy Truck Crankshaft Manufactured by Microalloyed Medium-Carbon Forging Steel. *J. Iron Steel Res. Int.* **2020**, *27*, 608–616. [CrossRef]
26. Wang, Q.; Cheng, G.; Li, J.; Dou, W.; Hu, X. Formation Mechanism of Large Inclusions in 80t 20Cr–8Ni Stainless Steel Casting for Nuclear Power. *Steel Res. Int.* **2019**, *90*, 1900349. [CrossRef]
27. Goto, K.; Matsuda, Y.; Sakamoto, K.; Sugimoto, Y. Basic Characteristics and Microstructure of High-Carbon High Speed Steel Rolls for Hot Rolling Mill. *ISIJ Int.* **1992**, *32*, 1184–1189. [CrossRef]
28. Lee, W.-H.; Liu, Y. Laboratory Evaluation of New Type of Backup Roll for Strip Shape Control. *ISIJ Int.* **2005**, *45*, 1636–1640. [CrossRef]
29. Prasad, M.S.; Ray, A.; Dhua, S.K.; Avtar, R.; Jha, S. Premature Failure of Work-Rolls in Tandem Mill: Some Microstructural Revelations. *J. Fail. Anal. Prev.* **2004**, *4*, 67–72. [CrossRef]
30. Kimura, T.; Ishii, M.; Amano, K.; Ueda, S.; Oka, Y.; Nakano, S. Secondary Hardening Characteristics and Those Effects on the Wear and Thermal Shock Resistance of 5%Cr-Mo-V Steel Roll for Cold Strip Mill. *ISIJ Int.* **1992**, *32*, 1224–1231. [CrossRef]
31. Parry, G.; Ostrovski, O. Wetting of Solid Iron, Nickel and Platinum by Liquid MnO–SiO2 and CaO–Al2O3–SiO2. *ISIJ Int.* **2009**, *49*, 788–795. [CrossRef]
32. Vantilt, S.; Coletti, B.; Blanpain, B.; Fransaer, J.; Wollants, P.; Sridhar, S. Observation of Inclusions in Manganese-Silicon Killed Steels at Steel-Gas and Steel-Slag Interfaces. *ISIJ Int.* **2004**, *44*, 1–10. [CrossRef]
33. Wang, K.; Jiang, M.; Wang, X.; Wang, Y.; Zhao, H.; Cao, Z. Formation Mechanism of SiO_2-Type Inclusions in Si-Mn-Killed Steel Wires Containing Limited Aluminum Content. *Metall. Mater. Trans. B* **2015**, *46*, 2198–2207. [CrossRef]
34. Shinozaki, N.; Echida, N.; Mukai, K.; Takahashi, Y.; Tanaka, Y. Wettability of Al_2O_3-MgO, ZrO_2-CaO, Al_2O_3-CaO Substrates with Molten Iron. *Tetsu-to-Hagane* **1994**, *80*, 748–753. [CrossRef]
35. Van Ende, M.-A.; Guo, M.; Proost, J.; Blanpain, B.; Wollants, P. Formation and Morphology of Al2O3 Inclusions at the Onset of Liquid Fe Deoxidation by Al Addition. *ISIJ Int.* **2011**, *51*, 27–34. [CrossRef]
36. Kapilashrami, E.; Sahajwalla, V.; Seetharaman, S. Investigation of the Wetting Characteristics of Liquid Iron on Mullite by Sessile Drop Technique. *ISIJ Int.* **2004**, *44*, 653–659. [CrossRef]
37. O'Malley, R.J. *Inclusion Evolution and Removal in Ladle Refining*; Missouri University of Science & Technology: Rolla, MO, USA, 2017.
38. Wu, S.; Guo, X.; Wang, Y.; Wu, G.; Lyu, S. Laboratory Study on Evolution Mechanism of Nonmetallic Inclusions in Al-Deoxidized Spring Steel. *Trans. Indian Inst. Met.* **2020**, *73*, 2807–2816. [CrossRef]
39. Kawakami, K.; Taniguchi, T.; Nakashima, K. Generation Mechanisms of Non-Metallic Inclusions in High-Cleanliness Steel. *Tetsu-to-Hagane* **2007**, *93*, 743–752. [CrossRef]
40. Deng, Z.; Zhu, M. Deoxidation Mechanism of Al-Killed Steel during Industrial Refining Process. *ISIJ Int.* **2014**, *54*, 1498–1506. [CrossRef]
41. Miao, Z.; Cheng, G.; Li, S.; Qiu, W.; Zeng, L.; Long, H. Formation Mechanism of Large-Size $CaO–Al_2O_3–MgO–SiO_2$ Inclusions in High Carbon Chromium Bearing Steel. *ISIJ Int.* **2021**, *61*, 2083–2091. [CrossRef]

MDPI
St. Alban-Anlage 66
4052 Basel
Switzerland
www.mdpi.com

Metals Editorial Office
E-mail: metals@mdpi.com
www.mdpi.com/journal/metals

www.ingramcontent.com/pod-product-compliance
Lightning Source LLC
Chambersburg PA
CBHW040716270926
51850CB00023B/2298
* 9 7 8 3 0 3 6 5 8 4 4 6 1 *